MW01283188

BLUEPRINTS

Also by Marcus du Sautoy

Around the World in Eighty Games

Thinking Better

How to Count to Infinity

The Creativity Code

What We Cannot Know

The Number Mysteries

Finding Moonshine

The Music of the Primes

BLUEPRINTS

HOW MATHEMATICS SHAPES CREATIVITY

MARCUS DU SAUTOY

BASIC BOOKS
New York

This book is dedicated with love to my children:
Tomer, Magaly and Ina

Basic Books
Hachette Book Group
1290 Avenue of the Americas, New York, NY 10104
www.basicbooks.com

Printed in Canada

Originally published in 2025 by 4th Estate, HarperCollins in UK
First US Edition: September 2025

Published by Basic Books, an imprint of Hachette Book Group, Inc. The Basic Books name and logo is a registered trademark of the Hachette Book Group.

Library of Congress Control Number: 2025931922

ISBNs: 9781541605695 (hardcover), 9781541605701 (ebook)

MRQ

10 9 8 7 6 5 4 3 2 1

Contents

Dramatis Personae
(in order of appearance)

Sometimes when you are about to embark on a grand adventure it is useful to have some idea of the people that you might meet on the way. The following list I hope will give the reader a helpful blueprint of the meetings and stories that they will expect to encounter through the book.

Olivier Messiaen, composer
Magicicada Septendecim, insect
Manfred Schroeder, architect
William Shakespeare, playwright
Radiohead, musicians
Björk, musician
Mark Haddon, writer
Carl Sagan, scientist and writer
Jorge Luis Borges, writer
Ahmes, mathematician
Giotto, artist
Eratosthenes, mathematician
Filippo Brunelleschi, architect
Antoni Gaudí, architect
Étienne-Louis Boullée, architect
Johann Sebastian Bach, composer
Charles Howard Hinton, mathematician
August Möbius, mathematician
Eugène Ionesco, writer
Le Corbusier, architect
Leonardo Pisano, Fibonacci, mathematician

Andrea Palladio, architect
Pythagoras, mathematician
Hippasus, mathematician
Carol Brown, choreographer
Hemachandra, poet and mathematician
Gregory Pincus, poet
Nitin Sawhney, musician
Sarngadeva, musician
Charles Darwin, scientist
Philip Glass, composer
Eleanor Catton, writer
the Nautilus, mollusc
Auguste Bravais, scientist
Luca Pacioli, artist
Leonardo da Vinci, artist and scientist
Béla Bartók, composer
Claude Debussy, composer
Amadeus Mozart, composer
Jackson Pollock, artist
Benoit B. Mandelbrot, mathematician
Helge von Koch, mathematician
Katsushika Hokusai, artist
Shah Jahan, emperor
Loren Carpenter, film-maker
Marvel Cinematic Universe, film-maker
Josquin des Prez, composer
Dmitri Shostakovich, composer
Tom Stoppard, playwright
Michael Frayn, playwright
David Foster Wallace, writer
James Joyce, writer
Rudolf von Laban, choreographer
Plato, philosopher
Archimedes, mathematician

Piero della Francesca, artist
Albrecht Dürer, artist
Johannes Kepler, scientist
Salvador Dalí, artist
Alex Garland, writer
Bernhard Riemann, mathematician
René Descartes, philosopher and mathematician
René Thom, mathematician
Zvi Hecker, architect
Johan-Otto von Spreckelsen, architect
William Hamilton, mathematician
Raymond Queneau, writer
Arnaut Daniel, troubadour
Thomas Mann, writer
Iannis Xenakis, composer and architect
Évariste Galois, mathematician
Richard Rhys, artist
Arnold Schoenberg, composer
Oulipo, writers and mathematicians
George Perec, writer
Italo Calvino, writer
Zaha Hadid, architect
Anish Kapoor, artist
Carl Friedrich Gauss, mathematician
János Bolyai, mathematician
Nikolai Lobachevsky, mathematician
M. C. Escher, artist
Henri Poincaré, mathematician
Emily Howard, composer
Christine Wertheim, artist
Margaret Wertheim, scientist
Fyodor Dostoevsky, writer
Jean Arp, artist
William Burroughs, writer

David Bowie, musician
Julio Cortázar, writer
B. S. Johnson, writer
John Cage, composer
Pierre Boulez, composer
Karlheinz Stockhausen, composer
Johann Philipp Kirnberger, composer
Carl Philipp Emanuel Bach, composer
Mark Wallinger, artist
Gerhard Richter, artist
AARON, AI artist

Overture

From the buildings we live in to the music we listen to, from the art we consume to the stories we tell, I am always blown away to discover that the structures that underpin human creation are invariably mathematical in nature. Time and again when I work with artists, creators and writers who are transforming the way we see and interact with our world, the structures they are tapping into for their creations are ones that I recognise as a mathematician. They seem to be drawn to the same ideas that thrill me, even if they have no mathematical background or technical experience in my world.

This might be surprising to those for whom mathematics was only ever a torturous exercise in multiplication tables. Art and mathematics. For many this would appear to be synonymous with chalk and cheese. A contradiction in terms. One the domain of emotional expression, passion and aesthetics. The other a world of steely logic, precision and truth. And yet scratch the surface of these stereotypes and one discovers that the two worlds have much more in common than one might expect.

The idea that connects the two domains for me is the concept of structure. Indeed, if I was going to define mathematics then I think 'the study of structure' is a good description. It is not the individual numbers but the structures that exist across all numbers that interest the mathematician. But these structures go far beyond just the numerical. Understanding abstract structure is what mathematics is all about. Because structure is an integral part of artistic practice, perhaps it is inevitable that there will be a connection.

In this book I want to explore some of the most fundamental mathematical structures that underpin human creativity. I have called these structures *blueprints*. They are, I believe, the hidden plans for the art, architecture, music and stories that humanity has created. Some of them are geometric, such as the circle or the Möbius strip, and can easily be understood as providing a model for how to make things. Others are numerical, such as prime numbers or Fibonacci numbers, which act like a code for building the world. But there are more conceptual structures, such as the ideas of randomness or symmetry, which are framing the way we think.

The idea of a blueprint was first introduced by the nineteenth-century polymath John Herschel into engineering and the construction industry. They acted like a negative of the original drawings with white lines mapping out the design on a blue background. The blueprint was a plan of the skeleton structure, which would then be used to create the physical object. The blue came from the photosensitive ferric compound that Herschel used. The idea here is that the mathematical structures explored in this book act as blueprints inspiring a whole range of different artistic constructions. The mathematics provides the plan or framework. The artist provides the flesh that realises the structure in the medium of their choice: words, paint, music or building materials.

One of the striking revelations is that it is not only human creation and innovation that taps into these mathematical blueprints. Nature too is exploiting these structures to achieve the wonders that the universe contains. It's as if human creativity and mathematical discovery are two languages with which to navigate and understand the physical universe we live in. The fact that we see these mathematical blueprints at work everywhere is powerful evidence for my belief that we live in a physicalised piece of mathematics. This book will explore the connections between these three points on a triangle: art, mathematics and nature.

Pieces of art have a moment in time when they emerge. That moment in history is integral to the work that transpires. The dynamism of the Baroque grows out of a society navigating a world in flux. The absurdity of Dada reflects the breakdown in political structures that led to war at the beginning of the twentieth century. The universe too seems to have had a beginning, which we can date to 14 billion years ago. The things we have discovered in this universe similarly have their own moment of creation: the first stars, the first cells, the first humans. Mathematics I believe has a different quality. I am a strong believer in the idea that the mathematical structures that underpin both art and nature are timeless. That they do not need a moment of creation. They exist outside time and space. The universe we live in is a physicalisation of these abstract structures.

There will be a moment when the human mathematician will see and articulate that structure for the first time in an act that often feels like a feat of creativity, but that moment is always mixed with a sense that this was a structure waiting there to be discovered, whose existence is independent of human involvement. Mathematicians are revealing the pure abstract structures that are the true origin of the things being projected onto the wall of Plato's cave. The universe we see around is simply a physical manifestation of those abstract forms. And within that universe artists are creating their own works which reinterpret these structures once again.

Artists often talk about this tension between creativity and discovery too. The statue was hidden inside the stone waiting for the sculptor to release it. This sequence of chords was waiting to be heard for the first time. I've talked to composers who admit to musical structures being discovered simultaneously by different musicians, much like the three nineteenth-century geometers who simultaneously discovered non-Euclidean geometry.

Although both mathematics and art might be uncovering the universal structures of the Platonic realm, Plato himself was in fact quite dismissive of art, believing it to be a poor copy of a poor copy. Mathematics was the real thing. Nature an approximation of the maths. And art a further degradation of the primary abstract structures from which it all evolved.

Plato has a point when it comes to nature failing in its attempt to physicalise these mathematical blueprints. We know the definition of a circle or a fractal or a prime number. And yet every attempt to identify a circle in the physical universe turns out to be a poor approximation of a true circle. At the atomic level, we see the pixelation that destroys the perfect nature of the circle's structure. The same for a fractal, where eventually we can no longer physically zoom in any closer without exposing the quantum pixelated nature of reality, while the mathematics carries on revealing ever smaller images of complexity. Prime numbers are infinite in nature and yet our finite universe can only ever represent an infinitely small proportion of these fundamental numbers.

Although Plato is right when it comes to nature's failure to match the perfection of mathematics, perhaps he is being too harsh when it comes to his view of the artist. In some ways, the artistic representations of these blueprints are succeeding at realising these structures where nature inevitably must fail. Although I am a Platonist at heart, the spirit of the book is to treat maths, art and nature as equal partners in our equilateral triangle of ideas.

Many of the creative artists that we will encounter are quite unaware of the mathematical structures that underpin their work. It is as if their aesthetic sensibilities and desire to experiment with form leads them to rediscover – and sometimes even to arrive first at – these mathematical blueprints. They might be heralded as artists but I want to honour them as secret mathematicians. I hope this book will reveal that we all have a bit of the secret mathematician in us, even if we outwardly might shun the subject. For those readers with a nervous mathematical disposition, I want to surprise you with the revelation that the things you love are often pieces of mathematics in disguise.

If mathematics acts as a powerful set of blueprints for human creativity, I also believe that an artistic mindset is an important blueprint for discovering new mathematics. This is a two-way dialogue, where mathematics and human artistic creativity are fuelling each other to reach ever greater heights. If artists are secret mathematicians, then equally mathematicians are secret artists. The mathematics itself might have an existence outside time, but it needs mathematicians to help the rest of humanity see these structures for the first time. The creation of mathematical structures is, I believe, driven by aesthetic sensibilities that resonate with the soul of the artist.

It's often forgotten, or never realised, that mathematics is a human activity, created by humans for humans, so it's designed to appeal to what makes us human. That is why it shares so much in common with the act of creating art. But what is often at the core of both of these activities is our relationship to the physical world around us. Both are creating ways to interpret, understand, navigate our place in the universe. And so it is entirely natural that mathematicians and artists should be excited, moved, intrigued, curious about the same fundamental structures.

The two cultures

One of the great tragedies of our modern age is that it regards art and science as worlds apart. Our education system often demands that we make a choice. Is it Rubens or relativity? Debussy or DNA? Shakespeare or the second law of thermodynamics? I found this insistence on compartmentalising my interests deeply frustrating. At school I fell in love with the power of science to help us navigate the world around us, to tell us where we've come from and, more excitingly, to look into the future and see what might happen next. But I also fell under the spell of the creative arts. Playing the trumpet in my local youth orchestra introduced me to the vast universe of music from Bach to Bartók, my local community theatre exposed me to the magical narratives of Shakespeare and Beckett, and despite never being able to express myself with a paintbrush I always enjoyed being transported by the artworks in the galleries we would explore on summer holidays in Italy or France.

I was rescued at school from this demand to pick a side when my mathematics teacher recommended a book about what it means to be a mathematician:

> A mathematician, like a painter or a poet, is a maker of patterns. If his patterns are more permanent than theirs, it is because they are made with *ideas* . . . The mathematician's patterns, like the painter's or the poet's, must be *beautiful*; the ideas, like the colours or the words, must fit together in a harmonious way. Beauty is the first test: there is no permanent place in this world for ugly mathematics . . . I am interested in mathematics only as a creative art.

The book was called *A Mathematician's Apology*, and it was written by the Cambridge mathematician G. H. Hardy. Reading Hardy's account of his life as a mathematician was a revelation for my 13-year-old self. The book provided a bridge between the two worlds of art and science that I had begun to fall in love with. On the one side I could see that mathematics was the language that science often resorted to in order to explain its ideas. On the other here was Hardy telling me that mathematics was a creative art comparable to poetry or painting. Indeed, Hardy seemed to be very dismissive of the parts of mathematics that were used in the sciences:

> These parts of mathematics are, on the whole, rather dull; they are just the parts which have least aesthetic value. The 'real' mathematics of the 'real' mathematicians, the mathematics of Fermat and Euler and Gauss and Abel and Riemann, is almost wholly 'useless' . . . It is not possible to justify the life of any genuine professional mathematician on the ground of the 'utility' of his work.

Most people's impression of mathematics is that it is a useful tool for solving scientific problems or building new technology, and that the principal motivation for discovering new mathematics is related to solving some practical problem. But if you ask most mathematicians about their research, you get a very different picture. The problems they are working on are very rarely motivated principally by utilitarian goals.

That's not to say that ultimately the discoveries might not have useful applications, but that is not the motivation for the time spent crafting these ideas. It turns out that aesthetic considerations very similar to those of the artists we shall encounter are the principal guiding lights which illuminate the direction that mathematicians choose to pursue.

Graham Greene wrote of G. H. Hardy's book 'I know no writing – except perhaps Henry James's introductory essays – which conveys so clearly and with such an absence of fuss the excitement of the creative artist'. It was a revelation for me to think of mathematics as something creative. But as I have spent more time in this world, I have begun to understand how much of a connection there is between creating art and creating mathematics.

It was G. H. Hardy's book that first opened my eyes to the creative side of mathematics. As well as the beautiful description that Hardy gave of being a mathematician, my copy of *A Mathematician's Apology* also included a fascinating foreword by his friend and colleague C. P. Snow. In it Snow recounted something of Hardy's life story and in particular his collaboration with the great Indian mathematician Srinivasa Ramanujan. This collaboration captured the importance of combining mathematical imagination and intuition, as embodied in how Ramanujan approached mathematical discovery, with the rigorous power of proof, which Hardy applied to confirm the truth of Ramanujan's insights. I enjoyed this part of the book as much as I did Hardy's *Apology* because Snow humanised what it means to be a mathematician by putting it in a historical context.

But C. P. Snow is perhaps best remembered for his 1959 essay *The Two Cultures*, in which he lamented the intellectual separation between the arts and humanities on one side and science and the natural world on the other. This divide, he believed, was a great handicap to the progress of human knowledge, but what Hardy's treatise impressed on me as a young teenager was how my own subject offered a bridge to span this divide. It is the eternal blueprints of mathematics that are at the heart of the way nature is constructed and which also frame our human responses to the natural world through our art.

Emotion

One of the unexpected bonds between the two is the role that emotion plays in guiding practitioners in both. For many people, art is the realm of emotion, mathematics the realm of logic. Art might have elements of mathematical structure embedded inside, but the principal attraction is its emotional content. Surely mathematics is guided by cold, clinical, logical argument? Emotion isn't part of proof.

What is perhaps a surprise for many non-mathematicians is that emotion is actually a strong guide for the mathematics we create. The choice of what we elevate to the mathematical canon and deem worthy of being published in our journals is a product of our emotional engagement with the ideas. Many people are under the impression that the goal of mathematics is to prove every true statement about numbers and geometry, as if we were trying to write all the books that it is possible to write or compose all possible pieces of music, creating a mathematical version of Borges's 'The Library of Babel'. What I've discovered is that mathematics is very different from simply a list of all the true statements we can discover about numbers. Mathematicians are storytellers. Our characters are numbers and geometries. Our narratives are the proofs we create about these characters. Not every story that it's possible to tell is worth telling.

One of my mathematical heroes, Henri Poincaré, said: 'To create consists precisely in not making useless combinations. Creation is discernment, choice . . . The sterile combinations do not even present themselves to the mind of the creator.' Mathematics, just like literature or music or art, is about making choices. And often those choices are made for similar aesthetic reasons to those that steer artists to choose what they create.

For Hardy, beauty was the first test for good mathematics. His talk of beauty in mathematics is commonplace among those who practise this art. But I think to just focus on aesthetics is to limit the impact of mathematics on the humans who engage with it. We want to be surprised by a moment of jeopardy, by a shift in the direction of the proof, to be provoked by an outrageous claim that is then resolved in a moment of revelation and acceptance of an arrival at a place where the world looks different. I always remember the moment Andrew Wiles breaks down in tears on camera at his recollections of the moment he understood how to prove Fermat's Last Theorem. Mathematicians experience a sense of awe and wonder at the revelations that emerge from spending time in this world. The structures that we create, inhabit, manipulate, explore are full of such surprises, moments of sheer excitement, of wonderment. The reaction to reading about these ideas for the first time is not far removed from hearing Bach's *St Matthew Passion* for the first time or feasting your eyes on a Leonardo painting or reading a Shakespeare sonnet.

The mathematics may have an abstract quality that means it seems far removed from our human experience. And yet the fact that the structures are at the heart of the way the universe works gives them an added sense of potency that is hard to ignore. This isn't just a satisfying sudoku that you've filled in which gives you a quick buzz of satisfaction. This is an insight into the structures that make the universe tick. Structures that might be the key to alternative universes. Or help you to see our own universe in a new way. Isn't this exactly what an artist is trying to do with their creative output?

When comparisons are made between mathematics and music, emotion is often used as an important distinguishing feature. As Leo Tolstoy is often credited as saying, 'music is the shorthand of emotion', and certainly there seems to be some sort of dictionary between musical and emotional expression. Shorthand is probably a good way of thinking about it because music is a lower dimensional structure when compared to the encoding of emotions in the brain. But what I feel gets missed in the tension between the emotional world of art and the logical world of mathematics is that very few artists talk about putting emotion into their work. Many composers would disagree that they are trying to describe their emotional worlds through this musical shorthand that Tolstoy talks about. The music is primary, the emotions are secondary.

For composers, emotion emerges as a consequence of the structures they have made. Our brain is reacting to the recognition of structure. The fascinating thing is how different structures elicit different emotional reactions. How that works in non-verbal domains like music is something of a mystery.

Stravinsky wrote about the fact that you don't write music to say something:

> For I consider that music is, by its very nature, essentially powerless to *express* anything at all, whether a feeling, an attitude of mind, a psychological mood, a phenomenon of nature, etc. *Expression* has never been an inherent property of music. That is by no means the purpose of its existence. If, as is nearly always the case, music appears to express something, this is only an illusion and not a reality. It is simply an additional attribute which, by tacit and inveterate agreement, we have lent it, thrust upon it, as a label, a convention – in short, an aspect which, unconsciously or by force of habit, we have come to confuse with its essential being.

Philip Glass says that he never deliberately programs any emotional content in his work. He believes it's generated spontaneously as a result of all the processes that he employs. 'I find that the music almost always has some emotional quality in it; it seems independent of my intentions.'

Very often in musical composition the structure comes first and the emotion emerges from that structure. And often that structure is innately mathematical in nature. As Stravinsky once said: 'The musician should find in mathematics a study as useful to him as the learning of another language is to a poet. Mathematics swims seductively just below the surface.'

It's certainly the case that people listening to music can have a very emotional response without any training in musical theory. But music is just a choice of notes of certain frequencies, with certain patterns that connect the notes across time. It is crystallised in black inked patterns on a white page. The point is that we have settled on patterns of notes that seem to correlate strongly with our emotional world. So these patterns, these structures that the composer creates, are perhaps an alternative encoding for the way our brains encode emotions in our neuronal and synaptic activity.

But if emotions are triggered by frequencies, and by patterns within these frequencies, might not the brain also be triggered by a more abstract recognition of the same patterns? Perhaps emotional reaction to art is precisely a response to the structures that resonate with the brain in a moment of recognition of something important and fundamental. Evolutionary pressures have very likely selected brains that become excited by the recognition of those structures which are the blueprints for the natural world, because it will give them an edge in navigating the universe around them.

Researchers at University College London using functional magnetic resonance imaging have found that, when mathematicians are exposed to mathematics they judge to be aesthetically pleasing, the same part of their brain lights up as that activated when people appreciate art or music. Professor Semir Zeki, who led the research, was interested in how often mathematicians talk about beauty in their work:

> To many of us mathematical formulae appear dry and inaccessible but to a mathematician an equation can embody the quintessence of beauty. The beauty of a formula may result from simplicity, symmetry, elegance or the expression of an immutable truth. For Plato, the abstract quality of mathematics expressed the ultimate pinnacle of beauty.

As Einstein put it: 'Pure mathematics is, in its way, the poetry of logical ideas.' The German mathematician Karl Weierstrass concurred: 'A mathematician who is not at the same time something of a poet will never be a full mathematician.'

Proof and truth

Part of the work I do as a mathematician is to try to create new structures. But another part is to tell the story of how and why the structures I have chosen to explore have an emotional resonance which makes them worthy of investigation. These stories are the proofs that we create. And this is often where the poetry of mathematics emerges.

G. H. Hardy, as part of his *Apology*, tells two such mathematical stories which gave me my first glimpse of the poetry that Einstein and Weierstrass were talking about. They are proofs discovered by the ancient Greeks and illustrate the drama and thrill of a good mathematical story. They certainly captured the imagination of my 13-year-old self. One was the proof that there are infinitely many prime numbers, which I will retell in my first blueprint. I remember how thrilling it was to read how the finite human mind can navigate the infinite through the finite manipulation of logical ideas.

The second proof was the revelation that there is no fraction whose square is 2. In other words, the square root of 2 is not a number you can write as a fraction. This was an extraordinary revelation at the time the discovery was made and still has implications today for the difference between the abstract world of mathematics and the physical universe we live in. This difference is a story that will crop up a number of times in our journey through the blueprints we will explore.

But there is an important role that proof plays in mathematics which perhaps marks it out as distinct from the act of artistic creativity. That's the power of proof to determine eternal truths. This ability to access truth is an important part of being a mathematician.

I still find it extraordinary that those 2,000-year-old proofs by the ancient Greeks which I read in Hardy's book are as true now as when they were first proved and still have such resonance today. Reading those proofs as an insecure teenager, that certainty was very appealing. No other discipline seems to have this sense of permanence. As Hardy himself wrote: 'Archimedes will be remembered when Aeschylus is forgotten, because languages die and mathematical ideas do not. "Immortality" may be a silly word, but probably a mathematician has the best chance of whatever it may mean.'

That desire for certainty is in stark contrast to the motivation that drives the artist. Ambiguity is a crucial characteristic of the creative arts that is anathema to those doing mathematics. Multiple interpretations, space to bring your own self to bear, your own act of creativity initiated by a work – all are totally valid and important aspects of the arts. Although mathematical creativity is meant to inspire new ways of looking at structure, and perhaps to initiate new discoveries in their turn by those who engage with your proof, ambiguity is not something that is encouraged in creating mathematics. You really want your readers to end up at the same destination as you by the end of the proof.

Although I chose to become a mathematician, a profession that many would regard as definitely on the scientific side of the 'two cultures' divide, the discovery that artists are drawn to very similar sorts of structures to those that excite me in my mathematical work has made me realise that we have much more in common than many appreciate. I've spent a lot of my time as a mathematician in dialogue with artists, discovering that what excites them is a mathematical structure that I recognise, just dressed up in a musical or linguistic guise. And, in turn, I've enjoyed sharing structures from my mathematical world that might inspire new artistic directions. These dialogues have very much informed the stories I want to share with you in this book. Mathematics, I've discovered, is a true bridge connecting the two sides of this cultural divide.

Most people recognise that mathematics is the language of science, that its equations and formulas have allowed us to understand and transform our environment. But what I want to do in this book is reveal that mathematics is at the heart of many artistic practices as well. That many artists, either consciously or often quite unknowingly, are tapping into structures that are part of the mathematician's cabinet of wonders. Mathematics is the science of structure. Music, literature, architecture and painting are the art of structure. Nature and science are the physicalisation of structure. These three realms – mathematics, art and nature – are united by their connection to these eternal blueprints.

Each chapter of this book is dedicated to a different mathematical blueprint. Primes. The circle. Fibonacci numbers. The golden ratio. Fractals. The Platonic solids. Symmetry. Hyperbolic geometry. Randomness. And every blueprint is a launchpad to exploring how the structure and its variations are used by a whole range of different artistic disciplines, as well as illustrating how nature too is underpinned by them. For those encountering a blueprint for the first time, I hope that engaging the structure through the prism of the arts might provide a way to connect with mathematical ideas that is less intimidating than meeting them straight on. But I will also use each blueprint to give a hint of the poetic and aesthetic side of mathematics.

In the end, what I have realised is that human culture is not divided into two camps, as our education system would have us believe: art on one side, science and nature on the other. It's really just one culture, with lots of different ways of looking at the world and languages for expressing it. The languages might be different, but the blueprints are the same. Whether we choose to become a mathematician, an artist, a musician or a writer, it is just about finding the right voice and language to express the story or idea that is burning inside us, bursting to get out.

One of the striking discoveries I've made in writing this book is how different artistic disciplines respond to the same structures. At first sight it might appear that some practices are more attuned to certain types of structure. Music and numbers. Architecture and geometry. Painting and proportion. But these structures resonate with whatever medium we create within. We see primes appearing in the literature of Shakespeare, dodecahedrons guiding dancers, Fibonacci numbers determining the proportions of buildings, music navigating the contours of negatively curved space. These structures are sufficiently abstract that they can be physicalised as sound, paint, music or dance.

It is with prime numbers that we begin our exploration of the blueprints that shape and create the world around us. The primes are in some sense a blueprint for everything: from primes we get numbers, from numbers we get mathematics, from mathematics we get the universe. So prime numbers are the beginning of it all.

Blueprint One
The Primes

On the evening of 15 January 1941, four prisoners carrying musical instruments emerged onto the stage in the barracks of Stalag VIII A. The temperature was −20 degrees Celsius outside. There was 50 centimetres of snow lying on the ground and all the windows were frosted over. The only heat came from the packed audience of prisoners and guards that had come to hear the concert. Word had got around that this was going to be the premiere of a composition by one of the famous inmates in the camp, the renowned French composer Olivier Messiaen.

Dressed in a tattered bottle-green uniform of a Czech soldier, which he'd been given after being stripped of his civilian clothes, and wooden clogs, Messiaen made his way to the piano that had been commandeered for the performance:

> . . . my God, what a piano! It was an upright piano with keys that stuck. So, when I played a trill it would stop and I had to pull the keys back up again in order to be able to go on. The violin was more or less standard, but the cello, alas, had only three strings. And the clarinet. Another catastrophe. One of the keys had been put near a heating-stove and had melted.

Messiaen apparently liked to exaggerate the state of the instruments that the quartet of musicians played on. The cellist later insisted that the cello did in fact have all four strings. 'You couldn't play the piece on three strings!' But Messiaen remembered the concert as a triumph against adversity.

The first challenge was to get the audience to be quiet. The 400-strong crowd were used to comedies and variety shows being performed, not delicate chamber music. Messiaen stood before them to say something about the composition they were about to hear. The piece was called the *Quartet for the End of Time*. This was not meant to be a play on words referring to the length of captivity, but to the end of past and future, marking the beginning of eternity. It was based on the text from the Book of Revelation in which St John describes an angel descending and declaring 'there will be no more time'.

In such apocalyptic times as Europe was witnessing in 1941, the sentiment resonated with prisoners waiting to hear the piece. As the musicians began to play, the clarinet and violin exchanged bird themes, one of Messiaen's big obsessions, and then Messiaen joined in with a sequence of chords on the piano.

I remember when I first listened to the opening movement, called the 'Liturgie de cristal', I had a strong sense of a pattern emerging in the piano part, but it was hard to pin down. The rhythm was a beautifully syncopated beat that repeated itself over and over. This was overlaid with a scrunchy sequence of what Messiaen called 'colour' chords, because his synaesthesia meant he saw colour whenever he heard certain combinations of notes. But when I focused on these chords rather than the rhythm, another pattern surfaced. They too seemed to be repeating after having run through a set sequence of notes.

Despite these patterns, the music never actually repeated itself, because the rhythm and the chord sequence always seemed to be out of step with each other. It was very difficult to hear what was going on. How was Messiaen creating this strange effect? Time is marked by seeing something repeat. The sun rising. The moon becoming full. The seasons coming round each year. But here was music where nothing quite repeated; almost, but not quite. Time was being held at bay, frozen, suspended.

In subsequent interviews with Messiaen and the other performers they recalled how the audience listened transfixed. Farmers and factory workers, doctors and priests, some who had never heard chamber music before, all were witnessing the beginning of a new sort of music. The rowdy rabble that had settled to hear the music now sat in silence for the hour of the performance. All trembled with the emotion coursing through the piece. It was as if the music was setting them free for that short hour. A moment frozen in time.

As the final notes faded away there was an awkward silence. Had they just witnessed a religious experience? Was it appropriate to clap? But then the applause burst out. 'Never before have I been listened to with such attention and understanding,' Messiaen later recalled. Everyone was aware that they had been part of something unique. The performance has gone down in the history of music as the most famous premiere of the twentieth century.

I remember being transfixed myself when I first heard the piece. I'd fallen in love with Messiaen's music in my late teenage years, after my youth orchestra spent a weekend playing through his explosive *Turangalîla-Symphony*. I think that weekend was the peak of my trumpet playing. It had to be. The trumpet part was crazily difficult. The whole piece was fiendish. Strange time signatures defied the usual three or four beats we were used to. But I remember being just blown away by the sheer ecstasy that Messiaen captured in the music.

That weekend spent inside Messiaen's sound world reshaped my musical landscape. It wasn't too long before I started exploring beyond that monumental piece and Messiaen quickly became my favourite composer for many years. It was during that period that I listened for the first time to the *Quartet for the End of Time* on a CD I borrowed from our local library. The delicate chamber work was of a different order of magnitude to the symphony we'd played, but I was equally transfixed.

For years I wasn't able to articulate the strange structure that I heard at work in that opening movement. How was Messiaen able to achieve this curious effect in the piano part where rhythm and harmony each individually repeated themselves in a regular pattern and yet, when they were put together, the pattern vanished. What was Messiaen's secret? It was only when I finally got a score that I was able to understand the blueprint Messiaen was exploiting. The answer turned out to be connected to one of my other teenage passions: prime numbers.

Prime numbers

2, 3, 5, 7, 11, 13, 17, . . . and off to infinity. These are the prime numbers. The indivisible numbers that cannot be written as two smaller numbers multiplied together. I first encountered these enigmatic numbers when I read G. H. Hardy's *A Mathematician's Apology*. Hardy himself had spent a lifetime obsessed with trying to unlock the secrets of these numbers and his book about being a mathematician uses their story as a way to access the passion for patterns that is at the heart of being a mathematician. Reading about these numbers quickly had me hooked too, as they have gripped generations of mathematicians before me.

First identified by the ancient Greeks, the prime numbers are the most important numbers in the whole of mathematics because all other numbers are made from these building blocks. Euclid's *Elements* gives the following argument to explain how every number is built by multiplying prime numbers together.

If a number isn't prime then that means it can be written as the product of two smaller numbers. But these numbers in turn are either prime or can themselves be divided again. Eventually you will hit a set of numbers that can no longer be divided. These are the primes that built the original number. Take, for example, the number 105. It is not prime, as it's divisible by 5: $105=5\times21$. The 5 is indivisible, a prime, but the 21 can be divided further: $21=3\times7$. But now we've got down to three indivisible numbers: $105=3\times5\times7$.

For a mathematician these numbers are like the atoms of arithmetic. The hydrogen and oxygen of the universe of numbers. Just as every molecule is made up of atoms from the periodic table, so every number is built by multiplying prime numbers together. Probably the most important discovery of chemistry was the patterns that Mendeleev uncovered that led to the creation of the periodic table. Every chemistry classroom and laboratory probably has an image of the periodic table on its wall, it is so fundamental to the subject. The 'period' in 'periodic' refers to the fact that atoms appear with similar properties at regular intervals in the periodic table. For example, a metal will often be followed by another metal after eight atoms in the table.

But despite 2,000 years of investigation, mathematicians have failed to reveal any pattern in the primes that might help us to understand when the next one will appear. Not only that, it turns out we can't even get away with writing the primes in a great big list that we can put up on the wall of the mathematician's lab. Because in one of the beautiful proofs created by the ancient Greeks, they revealed that the prime numbers go on for ever. There are infinitely many of the mathematician's atoms. There isn't a biggest indivisible number.

This might seem surprising, because at some point you might expect the numbers to get so big that they have to be divisible by something smaller. How can you be sure that beyond some point the numbers don't all become divisible? I am going to share this proof with you. It is one of the proofs which Hardy reproduced in his *Apology* that helped me fall in love with the beauty of mathematics. It doesn't involve any sophisticated mathematics beyond simple multiplication and division.

But the proof also reflects another important quality that I want to stress. That mathematical proofs have a narrative quality to them. Hardy chose this proof as a perfect illustration of the creative side of being a mathematician. These proofs are the mathematicians' musical scores, our poetry, our creations. Mathematics isn't simply a mind-numbing list of calculations and true statements about numbers. It involves choosing the things that are worthy of their place in the mathematical library. Mathematicians are storytellers. It's just our stories are about numbers and geometry. But those stories rely on tension and jeopardy, surprising twists and turns, and on emotional engagement by the reader as they are taken on the logical journey.

Reading a proof for the first time even for me as a mathematician is sometimes rather overwhelming, and I certainly don't always get it straight away, so don't feel disheartened if you don't understand everything. Give yourself up to the mathematical experience, just as those prisoners did when they listened to the very new music that Messiaen was writing. None of them would probably have chosen to attend a contemporary classical music event, and yet they were transfixed by what they heard even if they didn't understand every note. Reading this proof as a teenager did the same thing for me.

So here is the proof that there isn't a biggest prime number beyond which all other numbers become divisible. The story starts by making a classic move in the mathematician's playbook. Suppose the opposite. Suppose there does exist a biggest prime number. Let's give it a name: P. Now make a list of all the prime numbers from 2, 3, 5, all the way up to P. How do we show that actually there is a number bigger than P which is indivisible?

For me, this is like the moment Poirot reveals the identity of the murderer in one of Agatha Christie's stories and then explains why the culprit is responsible. The prime number culprit in our story? We multiply all the primes together from 2 up to P and then we add 1 to this to make a new number we shall call N:

$$N=2\times3\times5\times\ldots\times P+1.$$

If our list of primes 2, 3, 5, . . ., P were all the primes, then at least one of these primes has to divide the number N. Why? Because all numbers are either prime numbers or divisible by prime numbers as I explained previously. So which prime divides N?

The clever trick we've played, though, is to build a number that always leaves remainder 1 when we divide by any prime in our list 2, 3, 5, . . ., P. So it's not divisible by any of the primes! Does that mean N is a new prime number? If that were true it would be great, because it would give us a way to make new prime numbers. But this story is slightly subtler than that. What we have proved is that this number N is either prime or it is divisible by some primes that are not on our list. Whichever way, this means there are primes that are missing from the list. So P can't be the biggest prime number. There are more primes beyond P which divide this number N.

Even if you tried to claim that one of these missing primes is actually the biggest prime number, I can play the same trick again and produce yet another prime suspect. The prime numbers . . . they never run out.

The challenge now is to understand if there is any pattern or formula that can help to generate the prime numbers. There was a small moment when it looked like multiplying primes together and adding one to this number might be a formula for making primes. But this turns out to fail. For example, if you thought 13 was the biggest prime number and calculate 2×3×5×7×11×13+1 you get 30031. But 30031 is not prime because 30031=59×509. Building the number 30031 doesn't give you a new prime number but just reveals that we are missing the primes 59 and 509 from the list of primes up to 13. Our story only guarantees that there will always be primes missing from any finite list of primes we might write down.

For 2,000 years we've been trying to find a formula or some pattern to make sense of these wild numbers. But as you look through the list there just seems to be disorder and randomness. Don Zagier, a mathematician who has dedicated his life to uncovering the secrets of the primes, has had to admit:

> despite their simple definition and role as the building blocks of the natural numbers, the prime numbers . . . grow like weeds among the natural numbers, seeming to obey no other law than that of chance, and nobody can predict where the next one will sprout.

Prime numbers are the most enigmatic of characters and their story still remains unfinished. But what was their role in the performance that took place in Stalag VIII A on that freezing evening in 1941?

Prime number cogs

In the first movement of his piece, Messiaen wanted to create the strange sense of time ending. He achieved this in the most stunning manner. Time depends on things repeating, so he needed to produce a structure where you never truly hear the moment of repetition. While the clarinet imitates a blackbird and the violin a nightingale, the piano part plays a 17-note syncopated rhythm that just repeats itself over and over. But the chord sequence that the pianist plays set to this rhythm sequence consists of 29 chords, which are again repeated over and over. Such repeating patterns might lead to boredom and predictability, but not in this case. Because Messiaen's choice of numbers – 17 and 29 – means that something rather magical . . . or mathematical occurs. The numbers he chose are prime numbers and their mutual indivisibility means that the rhythm and harmony that Messiaen has set up never get back in synch once the piece is in motion.

The two musical ideas set off at the beginning of the piece, but as the 17-note rhythm sequences begins its second cycle, the harmonic sequence is still working its way through its 29 chords. It's only about 3/5ths through its sequence. Or, to be precise, 17/29ths of the way through. But once the harmonic sequence begins its second cycle, we are still five notes from finishing the second cycle of rhythm. The two musical ideas are kept out of synch. The choice of two prime numbers means that they don't get back in step until you have heard 17×29=493 chords, by which time the piece has already finished. You never actually get a moment of true repetition.

It's like two cogs, one with 17 teeth and the other with 29 teeth. As the interlocked cogs turn, they will have to go through 493 clicks before they simultaneously return to their original starting positions. Indeed, cogs in machines often are manufactured with a prime number of teeth precisely so that, as they turn, they will encounter different teeth on the cog they are engaging with, which evens out the wear of the teeth.

If you make different choices, for example an 18-note rhythm sequence against 30 chords, then after 90 chords you are back in synch because 90 is the smallest number divisible by 30 and 18. The different primes 17 and 29 keep the two out of synch so that the piece finishes before you ever hear the music repeat itself. This continually shifting and changing music creates for Messiaen the sense of timelessness that he was keen to establish. Although the music is moving forward, the fact that there is no true repetition makes the time strangely stand still.

Messiaen even adds another cog to this effect by having the cello repeat a 15-harmonic sequence against what the piano is doing. It has been calculated that it would take two hours for the three cogs to work their way through every position. The actual movement is over in under three minutes.

The idea of these cogs ticking as the piece progresses was the inspiration for an animation that I made with artist Simon Russell to explore the mathematics hiding inside the first movement of the *Quartet for the End of Time*. The animation begins inside a mathematical garden, but as the animation pans out as the piece progresses, you discover that this garden is in a fact a prison, alluding to the venue for its first performance. At the animation's heart, though, is this piece of clockwork with prime number cogs controlling the harmonic and rhythmic structure of the piece. I have performed the animation with live music and have always been blown away by how the musicians can perfectly keep in synch with the clockwork despite having their eyes glued to their musical scores.

Was Messiaen aware of the mathematical significance of these numbers? I asked composer George Benjamin, who worked with Messiaen in his later years, what mathematical training the composer had received. Benjamin thought he had essentially rediscovered the primes for himself. Playing around with numbers in his music had intuitively led him to this property that primes can keep things out of step.

It may be that Messiaen's interest in Indian music could actually have been the key to his attraction to prime numbers. Messiaen never visited India but instead stumbled upon techniques of Indian classical music while reading an entry on India in an encyclopedia. He was particularly taken by the extraordinary complexity of Indian rhythms, something I will return to in the blueprint devoted to Fibonacci numbers. As he explored these new structures, he realised that indivisibility was key to many of the interesting effects these rhythms produced:

> The indivisibility [Messiaen wrote] confers on them a sort of power that is very effective in the domain of rhythm. Among the Hindus, there are rhythms based on the number five (the number of fingers on each hand), the number seven, the number 11, all prime numbers.

Digging deeper into this world of Indian rhythms, he found even larger primes. A rhythm the Indians called the *lakskmica* was made up of 17 beats divided into 2+3+4+8. The *ragavardhana* rhythm was a 19-beat pattern made up of notes of length 2+3+2+12. He even found a rhythm called the *laya*, made up of 4+2+6+6+6+4+6+3 making a 37-beat pattern. The first two rhythms would form part of the ingredients for his *Quartet*.

As Messiaen played around with these rhythms, he understood how the primality meant that these structures just didn't fit neatly into conventional Western modes of music. But the tension they created was perfect for the music he wanted to conjure up. It seems to me he was aware that mathematics was the key to what he was doing. He would talk about the charm of impossibilities. 'This charm, at once voluptuous and contemplative, resides particularly in certain mathematical impossibilities in the modal and rhythmic domains.' The primes became Messiaen's blueprint for this charm of impossibilities.

Prime number cicadas

When there is a structure that emerges from both a mathematical perspective and an artistic one, invariably that structure has its origins in something that we will find in nature. Both mathematics and the arts emerge out of our encounters with the world around us. They are languages that we have developed to make sense of our environment.

But prime numbers are quite a sophisticated concept. At first sight they don't seem to be natural at all, and it is true that one does not see them all over the natural world as one does the Fibonacci numbers, for example. But there is a curious example of nature exploiting the property of primes which relates directly to the way Messiaen used them in the *Quartet for the End of Time*.

A few years ago, I took a trip to Nashville, Tennessee. Not for the country music but for a rather different sort of sound. A few days after my arrival, the normal hubbub of the city began to be drowned out by a high-pitched buzzing. The sound of summer in many countries around the world. Cicadas. Except in Nashville the sound was beginning to hit unpleasant levels. Trillions of cicadas had emerged en masse from the ground to fill the surrounding countryside. Half of them, the males, were singing their hearts out to court the females. The levels that have been recorded exceed 100 decibels, about the same volume as a passing ambulance. Bob Dylan's receipt of an honorary degree at Princeton University in 1970 was drowned out by the sound of a similar emergence of cicadas and is said to have inspired his song 'Day of the Locusts'.

Residents find the continual racket so unbearable that they often move out of town during the cicadas' emergence. The sole upside to the whole story is that they have to do this only once every 13 years, because this species of cicada has developed a very curious life cycle. They stay in the ground in the form of nymphs for 13 years, doing absolutely nothing except soaking up the sap from the trees. And then on this 13th year, they all emerge from the ground, almost all on the same day. That's when they take to the trees and start their party. They mate, they lay eggs and then after six weeks they all die and the forest goes quiet again for another 13 years before the next brood of cicadas emerges.

I talked to many residents in Nashville who could remember the last time the cicadas emerged. Some were still at school then but now were serving at the local Cracker Barrel where I was having breakfast. They described how the cicadas are completely harmless but that there is a dangerous moment when they all die. There are so many dead cicada carcasses strewn on the road that cars have to be careful not to start skidding on the river of insect bodies. One waitress told me of a website that couples who are planning a wedding can check to make sure that they aren't choosing a date when their vows will be drowned out by the singing cicadas.

But for me the excitement was about the number 13. A prime number. Was this just a coincidence? It seems not. In addition to 13 there are other broods across America that have a 17-year life cycle. The cicadas that accompanied Dylan's appearance in Princeton were examples of cicadas that only appear every 17 years. Across America you only find broods with these two prime number cycles: 13 and 17. There are no cicadas with a 12-, 14-, 15-, 16- or 18-year life cycle. It's curious, too, that you don't find these periodic cicadas anywhere outside North America.

One of the most striking facets of this story is how the cicadas manage to count up to 13 and 17. Nothing in nature's cycles mimics this periodic behaviour. The closest thing is sunspots, which repeat in a cycle of 11 years, but these would have no recognisable impact on a nymph in the ground.

One theory suggests that a cicada nymph is like a little bucket that, with each cycle of the seasons, is gradually filled by feeding on the sap of the trees. Once the bucket is full, this triggers the next stage in its life cycle and it appears out of the ground. Research at the University of California tested this hypothesis by altering the seasonal cycles of trees supporting cicada nymphs and did indeed induce premature metamorphosis of the associated cicadas. Cicadas that were in their 15th year were moved to peach trees that experienced two flowerings in a single year. Each flowering produces more sap to add to the cicada's bucket, so two flowerings in one year can trick the cicada into emerging a year early thinking that two years have passed.

But for me the real enigma is why they have evolved to have life cycles specifically with a prime number of years. The answer appears to be connected to the way Messiaen used the prime numbers in his *Quartet for the End of Time*. Primes are very good at keeping things out of synch. For Messiaen, it was rhythm and harmony. For the cicadas, there is a conjecture that the primes help them to avoid coinciding with a predator which also used to appear periodically in the forest, a predator that perhaps is not present in regions where the cicadas appear annually.

To illustrate this, imagine a predator which appears every six years in the forest. If the cicadas appear every nine years, not a prime number, then they will meet the predator after two cycles because 18 is divisible by 9, the cicada's cycle, and 6, the predator's cycle. Cicadas have no natural defences against predators bar safety in numbers. So they will not have had enough time to build up the size of their population, and they will end up getting wiped out. If, however, the cicadas appear every seven years, a prime number, then they can keep out of synch of the predator for longer. It takes 42 years for the two to coincide. 42 is the first number divisible by both 6 and 7. So although the seven-year cicadas are appearing more often in the forest than the nine-year cicadas, the indivisibility of 7 means they survive longer and can build up numbers to survive the attack of the predators that happens in the 42nd year.

It seems that in the forests of North America a real competition emerged between predator and cicada. Perhaps the predator changed its life cycle to try to get in synch with the cicadas. The cicadas would then have been forced to adapt and find a new prime number cycle. In the Nashville region it seems that 13 was effective enough to beat the competition, but in the forests around Princeton the cicadas had to extend as far as the next prime number, 17. It's not clear what this periodic predator was because its inability to find the primes probably led to its demise. Was it a poisonous wasp or even some deadly fungus? No one knows. The message is that if you know your maths, you survive in this world.

The cicadas still have predators, but the safety in numbers means that, as a species, they can cope. Today, when the cicadas emerge, the local bird population gorges on the huge supply of free food. Indeed, research has revealed that the caterpillars that the birds usually eat during non-cicada years get a free pass and themselves end up gorging on the leaves of the oak trees. But because the cicadas appear in such huge numbers, the birds fail to make a major impact on the population. If the cicadas spread their emergence over the intervening years, then there is a chance that they would get wiped out each season if their numbers were too low. They need to emerge en masse in this synchronised manner to survive.

It is striking to see the property of prime numbers emerging in such disparate contexts. One in a forest in Nashville, the other in a concert in a prisoner of war camp. The cicadas are to the predators as Messiaen's rhythm is to the harmony. Despite the very different scenarios, the primes are serving a common purpose: keeping things out of synch. The manifestations are poles apart, but the blueprint is the same.

Primes, though, are key not just to composing music and building up populations of cicadas. They also find a role in constructing the buildings we inhabit.

Acoustics

Architecture and prime numbers don't have an obvious resonance. The rather unnatural and awkward nature of primes doesn't make them the most obvious numbers to go to if you are designing spaces to live and work in. Other mathematical structures that we will encounter form much more natural blueprints for the buildings we've erected.

But there is one venue where prime numbers have played a crucial role in creating an architectural space that serves its function: the concert hall. If you're sitting listening to a performance of the *Quartet for the End of Time*, then it's important that your appreciation of the piece is not disrupted by the acoustics of the building. Too many flat surfaces and you start to get strange echoes. Too much soft furnishing and the sound just gets swallowed up. You also don't want your position in the hall to affect the balance of the music. If you're on the side that is located near to the piano, you still want to be able to hear the cello clearly.

I'm a massive Wagner fan and recently achieved one of the things I've had on my bucket list for some time: to hear the *Ring* cycle in the special theatre that Wagner built to stage his operas in Bayreuth. It is quite a remarkable space. For example, instead of the orchestra sitting in an open pit in front of the stage as in a conventional opera house, its players are hidden underneath the stage, out of sight. Traditionally, an opera singer has to belt out their arias in order to ensure they come through the wall of sound that the orchestra is making between them and the audience. But the effect of placing the orchestra underneath the singers meant that, in the performances I attended, the singer could be much more subtle in their delivery than I had ever experienced in any other opera house. I sat next to an old man for the four nights that I spent listening to the *Ring* cycle and it transpired that he had been a cellist in the orchestra for many years. 'You cannot believe how loud the orchestra sounds in that box beneath the stage,' he said. But the sound that we received was quite subdued.

The other innovation that Wagner initiated was the seats we sat on. He recognised that nice soft cushions were a disaster for the acoustics. They just swallowed all the sound. So comfort is out and hard benches are in. Given that Wagner operas start at 4 p.m. and end at 10 p.m., you need to be a pretty dedicated fan to endure the pain from the acoustically perfect but inordinately uncomfortable benches that Wagner had installed.

Actually, the Festspielhaus in Bayreuth isn't regarded as having the world's best acoustics. But Bayreuth does have another opera house in which Wagner considered staging his operas that does have wonderful acoustics. Completed in 1750, the Margravial Opera House is one of the few original Baroque theatres left standing in Europe. The Baroque style is all about ornamentation, and it turns out that all those Baroque frills act as rather good sound dispersers. For Wagner, however, it was just too small to accommodate the vast vision he had for his operas, especially given the size of the orchestras he employed.

The Grosser Musikvereinssaal, home to the Vienna Philharmonic and the venue for the traditional New Year's Day Concert from Vienna, is also celebrated for its acoustics. It was built in the late nineteenth century, and the ornamentation that was the style of the period is credited with diffusing the sound without absorbing it. The acoustics of the Grosser Musikvereinssaal and the Margravial Opera House were certainly achieved by chance rather than design. But they give some hints as to the key to good acoustics.

Three modern concert halls that are universally praised are Boston Symphony Hall, built in 1900 with a classic shoebox shape; Birmingham Symphony Hall, opened in 1991, which has surprisingly good acoustics given how many flat surfaces it houses; and the contemporary Philharmonie de Paris, which was inaugurated in 2015. This last hall is a masterpiece of cutting-edge acoustic technology.

But not every venue has been so successful acoustically. One particularly famous London venue, the Royal Albert Hall, opened by Queen Victoria in 1871, used to be notorious for its abysmal acoustics. The iconic dome, which makes the venue look a bit like an upturned jelly, unfortunately caused a strong echo that would bounce back on audiences.

One of the issues that has often materialised in concert halls built in the twentieth century is the fact that, although Modernist architects loved building spaces with smooth lines and curves in a minimalist and often brutalist fashion, these shapes are a disaster acoustically, reflecting back sound and creating terrible reverberation. The Royal Festival Hall, part of London's Southbank Centre built in 1951, is particularly problematic. The conductor Simon Rattle said of the acoustics: 'The Royal Festival Hall is the worst major concert arena in Europe. The will to live slips away in the first half hour of rehearsal.' I once played the trumpet during a concert in the hall with my youth orchestra in the 1980s and was unable to hear the oboist with whom I was meant to be performing a duet, despite the fact that she was just a few seats away.

Many concert halls like the Royal Festival Hall or the Royal Albert Hall have had to be reconfigured with new technology in order to correct their acoustics. And it was prime numbers that came to the rescue.

A purely reflective surface causes the sound to be received as a clear echo, and it will also be heard clearly in some regions of the space and not in others. What you want is a surface that diffuses the sound both in time and space.

Prime numbers turn out to be an effective way to achieve this. The discovery that a seemingly esoteric bit of mathematical number theory might help create good acoustics was made by the German physicist Manfred Schroeder in the 1980s. He suggested making a grid with elements of various heights, which would then be attached to the wall. In this way, the sound wave would be reflected at different times according to the relative heights of the pieces in the grid. The challenge was working out how the resulting waves would interact so that they didn't cancel each other out or cause strange reinforcements.

Once Schroeder untangled the equations, it turned out that prime numbers were perfect for determining the heights. It wasn't an obvious application for them, but depended on work that the German mathematician Carl Friedrich Gauss had done in the early nineteenth century on prime numbers and something called 'quadratic residues'.

If you take a clock face with a prime number of hours on it, for example seven, then you can use this to do calculations called 'modular arithmetic', or 'clock arithmetic'. The hours on the prime number clock run from 1 to 6 but then 7 is replaced by 0 because mathematicians find a 0 useful when doing arithmetic. If you want to add 4 to 5 on this clock, then rather than getting the answer 9, the clock hand moves 4 clicks on from 5 to land on the number 2. Since multiplication is just repeated addition it's also possible to calculate 4×4. You just add 4 together 4 times and discover that the clock hand goes from 4 to 1 to 5 to 2. So $4^2=2$ modulo 7. We say that 2 is a 'quadratic residue modulo 7' because it is the square of a number on this prime number calculator.

If you go through all the numbers on the clock face from 0 to 6 and calculate their squares, then you get the following sequence:

0, 1, 4, 2, 2, 4, 1.

What Schroeder discovered is that, if you use this sequence as the heights of the elements in the grid both horizontally and vertically, then the sound that bounces off the wall interferes in such a way as to create a good acoustic across the whole hall.

Larger prime numbers can also be used. For example the quadratic residues modulo 17 are:

0, 1, 2, 4, 9, 13, 15, 16, 16, 15, 13, 9, 8, 4, 2, 1.

Notice that there is always a symmetry in these numbers. That is because 16 is actually the same as -1 in these calculations and $(-X)^2=X^2$.

The fascination for mathematicians such as Gauss was the rather random distribution of these numbers, and that is actually what is being exploited by their use in acoustics. Despite the apparent randomness of these numbers, there are things you can prove about them. For example, for any choice of number N the numbers from 1 up to N will all be squares for some choice of prime number P. As a case of this general statement, for instance, the numbers from 1 to 6 occur as quadratic residues modulo the prime 71. Working out whether a number is a quadratic residue is so difficult computationally that it is used in modern cryptography.

Good acoustics are important not just for music but also for the spoken word. Theatres must pay attention to creating a space where everyone can hear every word. Shakespeare's Globe in London is a modern reconstruction of a wooden circular theatre – the very same 'wooden O' conjured up in the prologue to *Henry V*. The actor James Garnon, who has performed in many plays at the Globe, talks about the effect of this circular space encasing the players and audience: 'The acoustic is very, very true. It's like any wooden instrument. It requires subtlety and delicacy. If you attack it, you destroy the sound. You've got to tease the music out of it.' It's not clear how good the acoustics of the original Globe Theatre were. Primes might have helped modern acoustics, but things were pretty hit and miss in Shakespeare's day. However, that's not to say that the primes weren't helpful in building some of the effects Shakespeare wanted to create inside the Globe.

Shakespeare

Shakespeare is renowned as one of the greatest wordsmiths ever to have written in the English language. So when I was asked to do a presentation about 'Shakespeare and Maths' to celebrate the 400th anniversary of his death in 2016, I was initially rather stymied. I'd never thought of him as a great number cruncher. But then I had the good fortune one lunchtime to sit next to one of the English fellows in my college in Oxford, Will Poole. By the time we were eating dessert, I had had my ideas about Shakespeare blown out of the water by a series of fascinating insights that Will gave me. What might come as a surprise to many, and certainly did to me, is that Shakespeare loved his numbers. The hidden games that he played with numbers in his work reveal a sophisticated sensitivity to mathematics. Perhaps it isn't surprising, given that poetry depends on patterns in rhythm and rhyme.

It appears that within Shakespeare's circle of friends in London were a number of prominent Elizabethan mathematicians and scientists. One of the leading scientific luminaries of the day was the enigmatic John Dee, who was advisor to Queen Elizabeth. Dee was fascinated by everything from astrology to astronomy, alchemy to cryptography, magic to mathematics. It was a strange cocktail of science and the occult. He was arrested early on in his career for 'calculating'. Calculating per se was not illegal, but Dee was using his calculations to concoct horoscopes of the Queen, which was regarded as heresy. Shakespeare, even if he didn't actually meet Dee, was certainly aware of him by reputation. Some have even speculated that Dee was the inspiration for Shakespeare's creation of Prospero in *The Tempest*.

Mathematics in pre-Elizabethan England was regarded as having something rather satanic about it. The power of numbers to conjure the invisible led to its association with witchcraft, and mathematical texts were burned during Tudor times as books of magic and conjuring. Dee was one of the early advocates for rescuing mathematics from the fire and spoke up eloquently for its importance in a preface he wrote for the first English translation of Euclid's *Elements*. Published in 1570, the edition included beautiful pop-up polyhedrons to illustrate the geometric constructions.

That Dee's advocacy for the importance of mathematics was successful can be judged by the fact that many Elizabethan poets began to embed numbers of cosmological significance in their work. As the scientists of the era began to formulate models of the universe crafted out of numbers, poets too started experimenting with numbers as a blueprint for the self-contained worlds they were creating. But the freedom of the poet to create imaginary worlds allied them more to the mathematicians than the scientists of the time. As Julius Caesar Scaliger wrote in 1561 in his book on poetics: 'Poetry renders existing things more beautiful, and gives to non-existent things the appearance of existence; certainly it seems not to describe the things themselves, as an actor does (and the other sciences do) but, like another God, to create.'

In part, the poets of the day were reacting against Plato's belief articulated in the *Republic* that poetry was an inferior art form. In Plato's view, it got us further from the truth by being an imperfect copy of the imperfect world that was just a shadow of the real truth: the mathematically perfect Platonic universe. The Elizabethans were looking to capture this perfect mathematical universe and representing the truth of creation through the use of numbers in their poetry.

Edmund Spenser was a big fan of threading significant numbers through his poems to reflect the themes of the work. In his *Epithalamion* – an ode to his bride – he describes the day of the marriage and the year leading up to that day. To reflect this temporal journey, the poem is divided into 24 stanzas, one for each hour of the day. Among the poem's 433 lines are 365 longer ones, which mirror the days of the year leading up to their wedding. Spenser had to disrupt his rhyming scheme in the 15th stanza and misses out a line in order to create this representation of the year in his poem.

Shakespeare too liked to use his numbers to highlight themes. The numbering of his sonnets seems to deliberately reflect the themes of the poems. For example, Sonnet 60 opens with the lines:

> *Like as the waves make towards the pebbled shore,*
> *So do our minutes hasten to their end . . .*

Many scholars have suggested that it is no coincidence that the number 60 is key to counting the minutes in the hour. Sonnet 12 perhaps is even clearer in its connection between its theme and number. It opens: 'When I do count the clock that tells the time . . .' The number of the sonnet reflects the 12 hours of the clock face. And when Sonnet 52 refers to 'the long year', is the number of the sonnet actually a reference to the number of weeks that make up that year? This kind of numerological obsession for hunting hidden numbers in Shakespeare can lead you down a rabbit hole, but there is enough evidence to indicate that the numbers he was using weren't chosen by chance.

But, for me, the truly exciting discovery was the way that Shakespeare used prime numbers in his work. Shakespeare's rhythm of choice for much of his writing was the iambic pentameter. Iambic refers to the short pattern that gets repeated, in this case an unstressed syllable followed by a stressed syllable: dee-*dum*. The pentameter refers to the fact that there are five of these, making a total of 10 syllables in each line. The opening line of Sonnet 12 follows this pattern. Iambic pentameter had first been introduced to the English language in the fourteenth century by Geoffrey Chaucer. It certainly caught on. It is estimated that three quarters of all poetry written in English is composed using the same metre.

Some have suggested that the choice of 10 syllables for each line has its origins in the importance of 10 for the Pythagoreans of ancient Greece. The number was considered sacred to the sect, a number full of mysticism. This wasn't connected with our physiology – our 10 fingers and toes – but rather with a purer, geometric representation of the number. They would often represent 10 with a geometric pattern called 'the tetractys': a triangular figure of dots in four rows, one dot in the first row and four dots in the last, making a total of 10 dots.

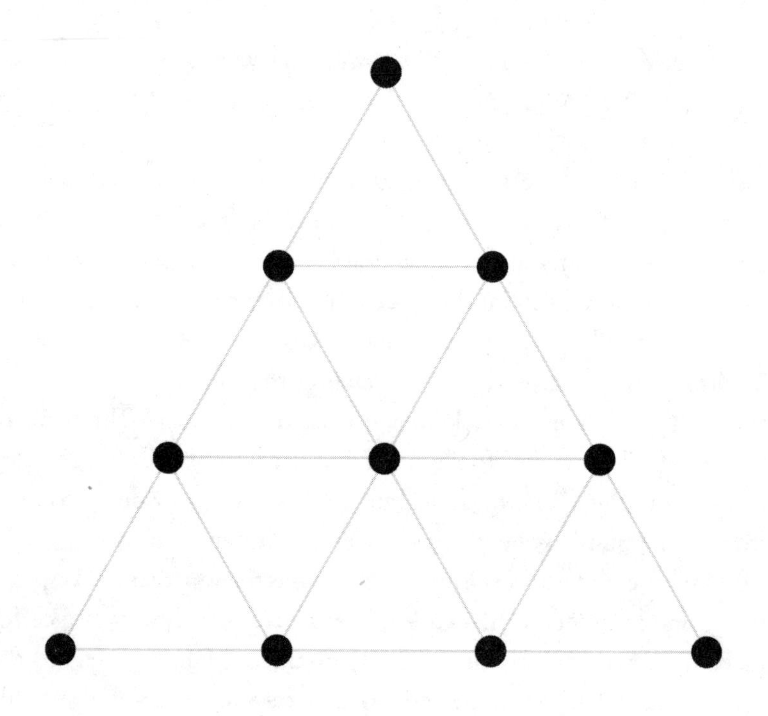

A number is called triangular if it counts the number of dots in triangles of arbitrary size. The importance of these numbers has even been suggested as the reason that Shakespeare chose to write 154 sonnets. In a slight fudge, the sonnets can be arranged into a triangle similar to the tetractys with 17 rows. The only trouble is that this accounts for 153 sonnets, the 17th triangular number, leaving those who propose this mathematical explanation having to account for the extra sonnet. Should the numbering start at 0 rather than 1? Is the 154th sonnet to be paired with the 153rd?

The 10s of iambic pentameter are the workhorse of Shakespeare's writing. But what is arguably the most famous line of Shakespeare?

To be, or not to be, that is the question . . .

Count the syllables and you find 11. The rhythm of iambic pentameter that runs throughout *Hamlet* can lull you into a soporific slumber. But this speech of Hamlet's is important. So Shakespeare wakes you up with that extra disruptive beat at the end of the line. Rather than a nicely divisible pattern of five lots of two, we get a prime number of beats. The indivisibility of the number when used in poetry makes the line stick out. The next three lines also have 11 beats. This use of an extra unstressed beat at the end of a line is known as a feminine ending.

One of the most famous lines in *Macbeth* also uses this prime number rhythm to alert the audience to a moment of significance as Macbeth contemplates the murder of Duncan:

Is this a dagger which I see before me?

But 11 isn't the only prime number that Shakespeare exploits. *A Midsummer Night's Dream* is one of my favourite of Shakespeare's plays partly because I played Flute the bellows-mender in a student production when I was studying in Oxford. We performed the play in the beautiful setting of the deer park of Magdalen College. My thespian passions were rather a distraction from the mathematics I was meant to be studying. But it was only some years later that I discovered that there was actually an overlap between the maths I was doing in the lecture theatre and the poetry we were reciting on stage.

Again the majority of the play is in the classic iambic pentameter. But listen to Puck as he rubs the magic potion into Lysander's eyes which will make him fall in love with the first thing he sees when he awakes:

Churl, upon thy eyes I throw
All the power this charm doth owe.
When thou wak'st, let love forbid
Sleep his seat on thy eyelid.

Seven syllables. For Shakespeare, whenever there is magic afoot, you'll find the number 7. It's like a code to indicate to the audience that something strange is happening in the play. The fairies in Titania's court also speak in seven-syllable phrases:

You spotted snakes with double tongue,
Thorny hedgehogs, be not seen.
Newts and blindworms, do no wrong,
Come not near our Fairy Queen.

In *Macbeth* it is the witches who talk in sevens:

When shall we three meet again?
In thunder, lightning, or in rain?
When the hurly burly's done,
When the battle's lost or won.

Macbeth refers to the witches as 'imperfect speakers', and this awkward seven is the key to creating an unsettling dialogue that feels like it doesn't quite complete itself. Ben Jonson compared the sound of the seven-syllable line to 'a brewer's cart upon the stones, hobbling'. It seems that even King James I, before whom *Macbeth* was first performed, was deeply unsettled by lines with an odd number of syllables. As he wrote in a treatise on poetry: 'Always take heed that the number of your feet [syllables] in every line be even, and not odd: as four, six, eight or ten, and not three, five, seven or nine.'

Although, on the whole, Shakespeare paid heed to the king's advice to stick to even numbers, there was one poem where he broke all the rules. I learnt about this poem not over a college lunch but on the football pitch while playing for the England Writers Football Team. One of my fellow defenders that day was Saul Frampton, another Shakespeare expert. While our strikers were doing their job at the other end of the field, Saul and I got talking Shakespeare and maths: 'You know his metaphysical poem "The Phoenix and the Turtle" is dripping with prime numbers?' In my excitement to find out more, we almost missed our opponents' counterattack: one of the occupational hazards of mixing football and literature.

This strange allegorical poem on the surface describes the funeral of the eponymous phoenix and turtle dove. It is attended by the other birds – the swan, the eagle, the crow. The mythical phoenix represents perfection. The turtle dove is a symbol of love.

So they lov'd, as love in twain
Had the essence but in one;
Two distincts, division none:
Number there in love was slain.

This curious last line hints at a potentially mathematical reading underpinning the poem. One suggestion is that the poem is alluding to the tension in mathematics which emerged from Euclid: that the concept of whole numbers and their ratios were not sufficient to measure the world. As I will explain in my third blueprint, the tension between geometry and number had revealed lengths that were irrational, that is, not expressible by the whole numbers. Is this what Shakespeare is referring to in the last line of the stanza, that 'Number . . . was slain'? The incompatibility of geometry and number is represented by the attempt to fuse in marriage the phoenix and the turtle dove. But in so doing both are extinguished:

Phoenix and the Turtle fled
In a mutual flame from hence.

To represent this incompatibility, Shakespeare exploited the unsettling power of prime numbers. The poem is made up of seven-syllable lines. It begins with 13 stanzas of four lines. It ends with what Shakespeare called a 'threnos' made up of five stanzas of three lines. The total number of lines is 67. Here we have the numbers 3, 5, 7, 13, 67. All prime numbers. Having seen that Shakespeare is a man who cared about number almost as much as the word, this choice of prime numbers cannot be considered accidental. They are there to represent the strange tension that runs throughout this curious poem.

As the beginning of the threnos states:

Beauty, truth, and rarity,
Grace in all simplicity,
Here enclos'd, in cinders lie.

It's as if the poem laments the discovery by the Pythagoreans that the universe is not made up of whole numbers, that it is imperfect, irrational, that the music of the spheres is full of discord. The poem talks of 'defunctive music'. The prime numbers help to sound out this disharmony.

Top of the primes

The power of primes to create discord has been taken up by more contemporary artists. Most Western pop music exploits the number 4. It is estimated that over 90 per cent of all pop songs ever written are in 4/4 time. So prevalent is the role of 4 in building pop music that 4/4 time is often referred to as 'common time' and indicated simply by a large C written at the beginning of the music. One of the explanations is that pop songs are for dancing and therefore even numbers work best for a species that dances on two feet. But some musicians enjoy disrupting those expectations.

The use of 5s in contrast to the conventional 4s is most famously demonstrated by the Dave Brubeck Quartet playing Paul Desmond's jazz standard 'Take Five'. First recorded in 1959, the five-note rhythm gives the piece that wonderfully laid back languid feel, as if it's not going to be rushed into moving along. The five made up of three plus two feels wonderfully natural in their hands, yet nonetheless that extra beat on top of the four catches one out each time. The Argentinian composer Lalo Schifrin, who wrote the theme tune for *Mission Impossible*, was led to a 5/4 rhythm after translating the initials MI into Morse code. MI becomes – – · ·, which if dashes are one and a half beats and dots are one beat makes five beats. He used to joke that the theme was actually written for mutant people with five legs to dance to.

If you want to create a sense of urgency in your music, then using a prime that misses a beat from a conventional grouping of four is very effective. Music in 7 or 11 beats does this very well because you are expecting 8 or 12, but then suddenly the music starts again one beat early. The Beatles used this trick in 'All You Need Is Love', where the phrases come in seven beats, creating a weird effect of the music starting a beat too early, as if your heart is skipping a beat from all that Love Love Love. Dionne Warwick's 'Say a Little Prayer' uses an 11-beat sequence, as does the bass riff in 'Whipping Post' by the Allman Brothers Band to generate something that has that wonderful urgency about it.

Radiohead are probably the recent band that have most consistently used all these tricks of strange time signatures and rhythms. Their use of prime numbers to disrupt expectations is responsible as much as anything for the band's unique and ground-breaking sound world. I defy you to listen to 'Everything in Its Right Place' and be able to count out what's going on. The name of the song is rather ironic given nothing seems in its right place rhythmically. The shifting back and forth between five beats and four beats and six beats creates the strange undulating sound that you can just sway to as if blown in the wind rather than dance to. 'Paranoid Android' does a similar thing, moving between eights and sevens, never allowing any one of the sequences to stabilise into a pattern, and giving the feeling every time you hear the seven beats that the music has glitched. It came as no surprise when I discovered that lead guitarist Jonny Greenwood, who composed much of

Radiohead's music, had, like me, fallen in love with Messiaen while playing in his local youth orchestra.

Björk probably pushed things further than any of her predecessors by using a repeating structure of 17 notes in her song 'Crystalline', part of her experimental seventh album, *Biophilia*. The song is meant to mimic the formation of a crystal and uses small musical units that build to make the greater whole. But it's the strange effect which this choice of 17 has that is most fascinating. When you listen, you are lulled into expecting things to reset after hearing four lots of four beats, making 16 beats. That's what conventional pop music would do. But Björk never likes to be conventional. That extra 17th beat almost acts to trip you up as you try to start again.

Prime time

Prime numbers have found a voice not just on the Elizabethan stage and in the modern recording studio but also in contemporary literature. If you open Mark Haddon's novel *A Curious Incident of the Dog in the Night-Time,* then you might initially be worried that your edition has some pages missing. The novel curiously starts at Chapter 2. But as you read on, the pattern, or lack of pattern, begins to reveal itself. The chapters are numbered using the prime numbers. There are no Chapters 4, 6, 8, 9 or 10. So what role are the primes playing for Haddon?

Rather than their indivisibility and their ability to keep things out of synch, as with Messiaen's use in the *Quartet for the End of Time*, or their disruptive rhythm, as used by Shakespeare in his poetry or Radiohead in their songs, Haddon is instead tapping into the character or personality of these numbers.

The narrator at the heart of the novel is Christopher, a 15-year-old boy with Asperger's syndrome living in Swindon, who describes his attempts to solve the mystery of who killed his neighbour's dog. Many people with Asperger's find social interaction very difficult because of the unpredictability of human nature. Instead Christopher finds comfort in the security and certainty that mathematics offers. Once you have proved something in mathematics, then you know with 100 per cent certainty that the statement is true. It won't suddenly make up its mind next week to be false.

Even the sciences can't offer this level of security. A theory popular today can be overturned by new experimental revelations of future generations. For Christopher, that security of knowing where you are with numbers is why they represent his safe space in a world of duplicity and ambiguity. For example, in moments of anxiety Christopher doubles 2s in his head to make himself feel calmer.

As Christopher writes in Chapter 101 (the 26th prime number): 'Mr Jeavons said that I like maths because it is safe.' His teacher explains that although a mathematical problem might initially look difficult, in the end when you get the answer, the logical explanation reveals that things are much less complicated than they first appeared. But Christopher realises that what his teacher is really saying is that life is not like maths: 'in life there are no straightforward answers at the end'.

I was struck when I saw Haddon's decision to use mathematics in this way because it was something that I recognised from my own experiences. Mathematics, for me, has always provided a safe place to escape the pressures of life. Ever since I was a kid, I've turned to the security of equations and theorems that wouldn't let me down. When the stress of growing up as a spotty teenager got too much for me, I used to recite the formula for the solution to a quadratic equation as a mantra in times of high anxiety. I resolved that it would be the last thing I uttered if I was ever on my deathbed. Christopher uses the very same formula for solving quadratic equations to cope with his trip from Swindon to London. Even in my adult life the mathematical world is one I escape to if I can't cope with the chaos and mayhem of my existence.

But there is something rather interesting about the decision to use prime numbers in the novel because currently they represent for us mathematicians a mystery, characters that we don't really understand. They are unsettling. We don't know when the next one is going to pop up. Their apparent randomness is deeply frustrating to us. And yet they also have a rigid quality that means 17 will always be a prime, no matter where or when you are in the universe. This curious contradictory nature – the randomness and the rigidity – is one of the aspects of these numbers that make them so fascinating.

And it is this dual nature that Haddon is exploiting when Christopher uses them to number the chapters of his account. They are like an unsolved mathematical murder mystery.

As Christopher explains in Chapter 19 (the eighth prime number): 'I think prime numbers are like life.' Christopher believes that when you have taken all the patterns away then it is the prime numbers that are left. He wrestles with the strange duality that prime numbers present. They are entirely logical and yet seem to obey no obvious rules. Even if there are rules that explain them Christopher thinks that you'll never find them 'even if you spent all your time thinking about them'.

Haddon isn't the only one to pick up on the personality of the primes as a theme to thread through a story. Prime numbers have a curious feature, that although they become rarer and rarer as you count higher and higher, we believe that infinitely often you'll see a pair of consecutive odd numbers where both are prime. Because of their close proximity, these pairs are called 'twin primes'. For example, 71 and 73, or 2,760,889,966,649 and 2,760,889,966,651.

The Italian author Paolo Giordano uses the second set of these two twin primes to represent the two characters at the heart of his novel *The Solitude of Prime Numbers*. Mattia and Alice both suffer traumas in their childhood, which results in them becoming outsiders. They meet as children and their close friendship continues into adulthood but never fulfils the romantic connection that bubbles underneath their relationship.

Mattia goes on to become a celebrated professor of mathematics and it is his musings on prime numbers that connect mathematics to the narrative of the novel. The primes are 'suspicious and solitary', we are told, 'which is why Mattia thought they were wonderful'. Through his description, we recognise that they have the same characteristics as Mattia and Alice. Outsiders. Lonely.

Except, as the narrator explains, there are these rather special primes that come in pairs: the twin primes. So close, yet frustratingly with an even number in between that prevents them from touching. They are destined to be isolated, lost in the silent universe of numbers, and yet here they both are, prime partners in the vast expanse of divisible numbers. Mattia feels that he and Alice are just such a pair of primes.

After one visit to Alice, Mattia returns home and writes '2760889966649' in the centre of a piece of paper. He speculates that 'no one else in the whole history of the world' had probably ever said this number out aloud. This is his number. And then, two lines below, he adds its twin, '2760889966651'. Her number. Destined to be paired and yet never touch. It is surely no coincidence that Giordano chose to tell his story in 47 chapters, a prime number.

I have a particular soft spot for twin primes partly because I am the father of twin daughters. I did try to persuade my wife to call them 41 and 43, but she baulked at the proposal. She's not so big on prime numbers. So these are my secret names for Ina and Magaly.

Twin primes have also been celebrated in verse. The slam poet Harry Baker often taps into his nerdy mathematical upbringing for inspiration. He studied mathematics at Bristol University, but he was also someone who always loved playing around with words. The two passions eventually collided in his poetry performances, and one of his most successful poems has twin primes at its heart.

59 has fallen in love with the girl next door: 60. But unfortunately the feelings aren't reciprocated:

> *While 59 admired 60's perfectly round figure,*
> *60 thought 59 was odd.*

59 eventually realises that they are just too incompatible to be together, but then he bumps into the girl next door to the girl next door: 61. Together they revel in their shared awkward primality. 59 and 61 – twin primes. A perfect match that leads to love.

You're 59, I'm 61, together we
Combine to become twice what 60 would ever be.

It is a beautifully poignant and yet wonderfully funny poem which helped Baker become Poetry World Slam Champion in 2012.

Primes: a universal language

Because mathematics is the blueprint for the construction of the universe, most science fiction writers believe that if the human species is going to communicate with alien life then mathematics will probably be the only shared language we'll have. That was certainly the belief of Carl Sagan when he wrote his novel *Contact*.

Ellie Arroway, the central character in the book, works for the organisation SETI, the Search for Extra Terrestrial Intelligence. This is a real initiative in which scientists monitor the electromagnetic radiation arriving on earth for any possible signals that might indicate intelligent life is trying to contact us. I had the chance to visit the SETI Institute in California some years ago and it was fascinating to talk to scientists who were pinning their hopes on finding a radio needle in this cosmic haystack. Although they recognised how crazy the project might sound, finding a signal would be a game-changer for our place in the universe.

Nikola Tesla was one of the first to contemplate the idea that we might detect signals from intelligent life. His target was Mars. He even thought he'd picked up a message in 1899 when his equipment recorded a strange repetitive signal that cut out when Mars sank below the horizon. No conclusive explanation has ever been given for the origins of the signal.

Carl Sagan had been fascinated for decades by the idea of life existing beyond our planet. In 1972, with his friend and fellow astrophysicist Frank Drake, he designed the plaque that was attached to the side of the Pioneer 10 spacecraft. The plaque was intended to provide a message about the species that had created it, in case it was picked up by alien life once it had left our solar system. They faced the interesting challenge of how to encode messages on the plaque so that an intelligence that didn't speak English, or any other human language, might understand. Their solution was a combination of mathematics, science and an appeal to the universal power of a diagram or picture.

The idea of how aliens might choose to contact us was very much on Sagan's mind as he started writing *Contact* in the early 1980s. Ellie Arroway spends her days in New Mexico listening to the background static of the universe, desperate to hear something that is more than just noise. Her colleagues are very dismissive of her wasting so much research time on such a seemingly futile endeavour.

But then Arroway picks up a beat. Stars such as pulsars can emit regular pulses like a drumbeat, but this is different. The number of beats in each burst changes. As Arroway begins to interpret the sequence of beats in each burst she starts to spot a pattern, or rather a significant lack of pattern. She first has to convert the beats from binary into decimal, but she's okay doing that in her head. Once in decimal, the results are unmistakable: 59, 61, 67, 71. Prime numbers.

> A little buzz of excitement circulated through the control room. Ellie's own face momentarily revealed a flutter of something deeply felt, but this was quickly replaced by a sobriety, a fear of being carried away, an apprehension about appearing foolish, unscientific.

The prime numbers persist all the way up to 907, and then the next burst resets itself and begins again from the smallest prime. 155 prime numbers. No errors. No omissions. Arroway knows that this is significant.

One of the reasons that Sagan chose the primes rather than the Fibonacci numbers, for example, is that we know of no natural process that generates this strange enigmatic sequence. We saw the cicadas exploiting individual primes for their evolutionary survival, but it is striking that the primes are exceedingly rare in nature despite being the building blocks of all numbers. The Fibonacci numbers, in contrast, are all over the natural world. Receiving these in a message would not have been so significant. They might easily have been generated by some unintelligent natural process.

For Sagan, a list of primes encoded like this in electromagnetic radiation could only have its source in an intelligent life form. The universality of the primes across the cosmos also makes them good candidates for a greeting that is likely to be recognised by another species. The number 59 is as indivisible in the star system Vega as it is here on earth. The alien life form might have a different chemistry, biology, even physics, but its maths will be the same. Mathematics does not depend on a physicalisation to manifest itself unlike the other sciences. It is sitting there as a set of structures ready to be the blueprints even for other universes beyond our own if the multiverse turns out to be a reality.

In our own attempts to communicate with life outside the solar system, primes have also been an important ingredient. In 1974, to celebrate the refurbishment of the Arecibo Telescope in Puerto Rico, it was decided to transmit a message towards a dense collection of stars called M13. There's a good chance one of the stars in this stellar cluster has a habitable planet orbiting it. The only trouble is that M13 is 25,000 light years away, so they won't have got our greeting yet. And a response? That will take another 25,000 years. Not the snappiest of exchanges.

Once again, the message was designed by Sagan and Drake. Like the message received by Ellie Arroway in *Contact*, the broadcast sent from the Arecibo Telescope consisted of an extended radio burst modulated between two frequencies. The challenge for the alien civilisation would be to decode the message contained inside this seemingly random-sounding signal. The first step would be to realise that the repetition of only two frequencies meant that this was something written in binary. Perhaps a sequence of numbers like the ones Arroway detects.

However, the message on this occasion wasn't mathematical, but visual. The higher frequency was meant to be interpreted as a bright pixel and the lower frequency as a dark pixel. Received as a long stream of bright and dark pixels, at first a picture would be unlikely to emerge.

The hope was that a bright alien mathematician would spot the significance of the number of beats: 1,679. This is not a prime number but

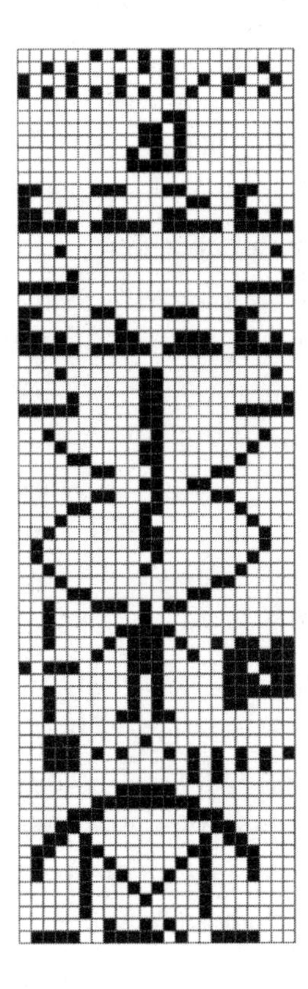

the product of two prime numbers: 1,679=73×23.

One of the important facts about prime numbers is that there is no other way to make 1,679 by multiplying different primes together. Writing 1,679 as 73×23 is the key to constructing the picture contained within the Arecibo message. The aliens will need to lay the pixels out in a 73 by 23 array. There are two different ways to do this: either 73 rows of 23 pixels, or 23 rows of 73 pixels. It's the first option, though, which produces an image that seems to consist of identifiable patterns.

It is interesting to see if an average human can successfully decode what is being communicated in this image. Probably the only part that everyone would recognise is the picture of us in the middle. But weirdly that would probably be the last part that aliens would be able to identify. The rest is an attempt to communicate more universal ideas that would be common to both of us.

The first four rows are the numbers from 1 to 10 in binary: 1, 10, 11, 100, 101, 110, 111, 1000, 1001 and 1010. Some interesting decisions have been made over how to represent the numbers. One pixel is being used at the bottom of the number like a little plinth for it to sit on. Probably because of the constraints of the 73 by 23 format, when we get to representing 8 as 1000, part of the number shifts to a new column. Not the easiest to interpret and we've only just counted to 10.

From there we move on to the atomic numbers for the atoms that make up DNA: hydrogen (1), carbon (6), nitrogen (7), oxygen (8) and phosphorus (15). This is a clever mix of the universal and the particular. The periodic table of atomic elements is probably the same across our local neighbourhood of the universe, so that any nearby alien life will have made the same list in the same order. But the particular molecule made up of these five elements might well be very specific to our biology.

The message continues with more information about the composition of DNA, concluding with the pixelated picture of the human body. The message then tries to give some clue as to the source of the signal. A pixelated diagram of the planets in our solar system precedes a picture of the Arecibo Telescope.

My hunch is that any aliens who received the message would be able to make sense only of the maths. DNA might be the blueprint for our species but that was probably the result of some random combination of factors. Even around the corner from us in the star system M13, a completely different form of life might have emerged from the chemical soup. But if that life form is able to count, then 17 is going to be as prime there as it is here on earth.

The primes are the beginning of the contact described in Sagan's novel, but ultimately its story concludes with a message hidden inside another universal structure: π. This number is key to our next blueprint: the circle.

Blueprint Two
The Circle

A magician arrives by bamboo canoe from the infinite villages upstream and heads to the circular ruins that were once a temple. He has no memory of his life before he arrived by boat, but he has a goal: to dream a man and insert him into reality. Beneath a full moon whose circular disc is perfect, the magician dreams a beating heart. It takes a year of dreaming for this heart to grow to a complete man ready to become his apprentice.

The god of the temple, whose name is Fire, appears in his dreams and instructs him to send his creature downstream. But before he sends his creation on its journey, so that his prodigy will not be aware that it is a phantom, he erases the memory of its years as apprentice to its creator.

After his prodigy departs for the ruins downstream, fire consumes the magician's temple, but the flames do not burn him. 'With relief, with humiliation, with terror, he understood that he too was a mere appearance, dreamt by another.'

So ends 'The Circular Ruins', one of the most famous short stories of the Argentinian writer Jorge Luis Borges. The circle is a shape that runs through many of this writer's enigmatic tales. Despite being celebrated today as a master of the form, rather curiously he had never intended to be a writer of short stories.

It was an accident that Borges suffered on Christmas Eve 1938 which was responsible for his first experiments in the genre. Borges had begun his literary life writing poetry and publishing articles in literary journals, including *El Hogar*, an Argentinian version of the *Ladies' Home Journal*. But on that fateful Christmas Eve, he struck his head on an open window when he was running upstairs to meet a lover. The sepsis that he suffered because of the wound almost killed him. To test whether his brain was still working, Borges decided to experiment with writing short stories as a new challenge.

The stories that resulted forged Borges's name as one of the great writers of the twentieth century. He never won the Nobel Prize for Literature, but Borges was always quite sanguine about the snub: 'Not granting me the Nobel Prize has become a Scandinavian tradition; since I was born, they have not been granting it to me.'

I fell in love with Borges's world after my French mathematical collaborator François Loeser introduced me to his stories. After a long day battling with motivic integration at the École Normale Supérieure, we retired to a local Parisian café to share some of our interests outside mathematics. We turned out to both be fans of Tintin. One of his other passions was the stories of Borges. When I looked puzzled, he replied: 'What! You haven't read Borges? As a mathematician, you just have to read Borges!' He was right. I haven't looked back, and he soon became one of my favourite authors.

Borges's short stories are a wonderful literary exploration of the ideas of infinity, of paradox, of the nature and shape of space, and, in particular, the circle. He was a voracious reader and he spent his childhood plundering the books in his father's library. He talked about dedicating himself more to reading than living. In 1955, he was appointed director of the National Public Library in Buenos Aires. The job should have brought him much joy, but he very sadly went blind that year; an ironic twist that could easily have been a plot development in one of his short stories. As well as books of literature, poetry and philosophy – especially the works of Bertrand Russell and Arthur Schopenhauer – Borges also encountered texts explaining the new ideas of science and mathematics that were emerging in the twentieth century.

One of the books he became entranced by was *Mathematics and the Imagination* by Edward Kasner and James Newman. In his review of it for the literary journal *Sur*, he predicted that it would join the select collection of books that he repeatedly reread and scribbled with notes. He made a list of the mathematical ideas that excited him: 'the fourth dimension . . . the mildly obscene Moebius strip . . . transfinite numbers, the eight paradoxes of Zeno, the parallel lines of Desargues that intersect in infinity, the binary notation Leibniz discovered in the diagrams of the *I Ching*, the beautiful Euclidean demonstration of the stellar infinity of prime numbers'. The authors of the book passionately tried to communicate these heady ideas to those not equipped with the technical language of mathematics to understand them. You can feel the thrill instilled in Borges as he read.

Among the topics explored in Kasner and Newman's book was the work of the German mathematician Georg Cantor, who had produced exciting new ideas about infinity at the end of the nineteenth century. Before then, infinity had represented the unknowable. But Cantor came up with ideas to tame it, to understand that there wasn't just one infinity but many, some bigger than others. He used the Hebrew letter *aleph* to codify these infinities. I remember being blown away by Cantor's ideas, just as Borges was, when I first read about them. His theorems about different sorts of infinity rank in my top ten theorems to take on a desert island.

Borges's fascination with these ideas led to one of his great short stories, called 'The Aleph'. In a house in Buenos Aires, you can see the Aleph if you focus on the 19th step of the stairs down to the basement. The Aleph is a point in space that contains all other points. Anyone who gazes into it can see everything in the universe from every angle simultaneously, without distortion, overlapping or confusion. It is infinity in a point. It is no wonder that Borges opened the short story with Hamlet's words: 'I could be bounded in a nutshell, and count myself a King of infinite space.'

As well as new ideas of infinity, the public was wrestling with the strange science of quantum physics and Einstein's theories of relativity that had emerged at the beginning of the twentieth century. These theories described new hypotheses of time and space requiring geometry with more than three dimensions, another idea explored in *Mathematics and the Imagination*. The challenge of the infinity of time was one that particularly intrigued scientists. If there was a 'big bang' that started our universe as Einstein's theories predicted, what came before that? Did it make sense even to ask such a question? And what was the ultimate destination of our own cosmos? An infinite future, or one that might result in a new beginning? As Borges sent his magician downstream in 'The Circular Ruins', these were the themes that he was exploring. Without the technical mathematical and scientific knowledge to explore these ideas, Borges resorted to the tools that he was so well equipped to wield: narrative and language.

The idea of the circle as a narrative device was one that particularly appealed to Borges. Narrative traditionally was linear in structure: a beginning, a middle and an end. But writers in Borges's generation were beginning to experiment with the idea of circular narratives, mimicking the emerging scientific idea of cycles of time and space. James Joyce's *Finnegans Wake* is constructed as a circle as it ends in the middle of a sentence which one realises is finished by the half sentence that begins the whole book.

For Borges, the circular story at the heart of 'The Circular Ruins' was a model for the possible circular story of our universe. Is our universe just the dream of a previous universe, heading back upstream through the infinity of time? In an essay that Borges wrote in 1936 called 'The Doctrine of Cycles', he talked about the impact that a finite universe would have on time. He argued that since we believe that the number of atoms in the universe, although very large, is nonetheless finite, this implies there is in turn only a finite, if also large, number of permutations of these atoms. Therefore given an infinite amount of time, the universe will exhaust all these permutations and hence will repeat itself.

> Once again you will be born of the womb, once again your skeleton will grow, once again this page will reach your same hands, once again you will live all the hours until the hour of your incredible death.

The science that Borges read about implies that a finite universe is destined to repeat itself, just as he described in 'The Circular Ruins'. The importance of the circle as a blueprint not just for the universe but also for the creations of the human species can be measured by the fact that this shape is responsible for some of the earliest mathematics and buildings made by humans.

The circle

If the primes are the building blocks of numbers, then the circle probably represents the beginning of geometry. The circle is first defined formally in Euclid's *Elements*. In Book 1, Euclid wrote:

> A circle is a plane figure bounded by one curved line, and such that all straight lines drawn from a certain point within it to the bounding line are equal. The bounding line is called its circumference, and the point its centre.

The *Elements* dedicates many of its pages to the importance of geometry. And it is indicative of the importance of the circle as a fundamental shape to geometry that it appears in the first of the 13 books. One of the themes that runs throughout the *Elements* is the idea of shapes that you can create with the tools of a straight edge and compass. The line and the circle are the building blocks of all shapes in the *Elements*. A circle can be created simply by attaching one end of a rope to a stick in the ground and then, keeping the rope taut, allowing the other end to map out a shape equidistant from the fixed point.

The ease of making a circle is why we see this shape appearing in the very first constructions that humans made on this earth. Ancient stone circles created by Neolithic cultures 5,000 years ago are scattered across north-western Europe. They vary in size from the impressive edifice of Stonehenge to the delicate circle of stones found in Swinside in Cumbria. Some of these circles are on such a scale that it is hard to think quite how the makers of these shapes achieved the designs. At Avebury, in Wiltshire, the stones are so widely spaced that the outer circle of stones runs for over a kilometre. To build such circles would have required sophisticated mathematical skills.

What was the significance of the circle for these early cultures? To understand the answer to this it is instructive to think where early humans would have naturally encountered the circle. As Borges alludes in his story of 'The Circular Ruins', the full moon would have been a regular circle that humans would have encountered in the night sky. The sun too would have been seen as another important circle, especially as it set in the evenings and could be stared upon in its full circular glory. On a more earthly scale, the rainbow is part of a circle, the ripples of water after a stone hits the surface would also have allowed humans to make transient circles. Even staring into the eye of another being we would have encountered the iris, the circle at the centre of the eye, a circle that might represent a window into the soul. The circle was there for all to see as a key blueprint for the way nature was constructed.

The dynamic nature of the moon's shape, which only achieved a complete circle every 29.5 days, must have had a mystical significance for early humans. It certainly was important enough that this fluctuating circle was a significant measure of time for many cultures. The passage of time is another place where early cultures would have experienced a more abstract notion of the circle. Time can only be measured when we see the repetition of some phenomenon. The sun rising again each day, the moon cycling through the monthly phases, the seasons changing over the course of a year. The physical circle became a symbol in many cultures for the cycles of time.

These celestial manifestations of the circle led to the symbol being associated with the heavens, with the gods, with the act of creation. In the art of ancient Egypt, the circle features prominently. The god Ra is depicted as a human with the head of a falcon crowned with a circle representing the sun. The first representation of the circular ouroboros, the image of the snake eating its own tail, can be found in the tomb of

Tutankhamun. This circle has become a recurring image for the idea of an eternal cycle of destruction and rebirth throughout history.

The circle is of fundamental importance in Buddhist and Hindu cultures. When Shiva is represented as Nataraja, the God of the Dance, he is always dancing inside a circle of fire, which in Hindu cosmology represents the creation and destruction of everything in the never-ending cycle of time. The mandala is an important symbol of the spiritual journey that human life embarks on and is used as an aid for meditation. The word *mandala* is Sanskrit for 'circle', and these geometric figures are often created within the boundaries of a circle. The circle is also used to depict those who have achieved enlightenment, where a halo is drawn above the head of the enlightened one. This idea of a holiness circle is picked up by early Christian depictions of saints and other biblical figures.

The ancient Egyptians were among the first to tease out some of the important mathematical properties of the circle and in particular to understand that the geometric shape encoded one of the most important numbers in mathematics: π. Their mathematical analysis grew out of an important natural occurrence of the circle in the meandering of the River Nile. Rather than a straight line, the course of the Nile carved out arcs that approximated to parts of the circle. Egyptian civilisation had reached a stage where its rulers wanted to tax people for the land they owned. Most areas of land were bounded by simple rectangles whose area was simple to work out. But those bordering the Nile were more challenging shapes to calculate.

Measuring the half circles of land carved out by the River Nile led to the discovery that a circular area of land was always a constant multiple of the square of the radius of the circle. This multiplying factor is the number π. The area of a circle, however large or small, is always πR^2, where R is the radius of the circle. In one of the first significant mathematical documents in the history of mathematics, the Rhind Papyrus, we find the first attempts to use the circle to calculate π. Written by an Egyptian scribe called Ahmes in about 1550 BCE, the papyrus is housed in the British Museum, and it is full of fantastic mathematics, including the first estimate for a value for π.

Ahmes tried to estimate the area of a circular field whose diameter is 9 units across. Because the area of a circle is π times the radius squared, if we know the area A and we know the radius R we can calculate $\pi = A/R^2$. The Rhind Papyrus states that a circular field with a diameter of 9 units is very nearly equal in area to a square with sides of 8. But because the area of the square is easy to calculate, that provides a way to estimate the area A of the circle. The radius R in this case is half the diameter, namely 9/2 units. So Ahmes' calculation gives the first estimate for π as 64 (the area) divided by 81/4 (the radius 9/2 squared).

This comes out at approximately 3.16. Today we know that the decimal expansion of π starts 3.14159, so this is not bad for a first estimate. As mathematics developed, so more cultures had a go at trying to capture this important number.

Like Ahmes, the ancient Greek mathematician Archimedes, in an essay written around 250 BCE, also tried to capture the circle using other shapes that he was able to analyse more easily. Archimedes started by drawing a triangle inside and outside the circle. The triangle doesn't look much like the circle. But what if he now doubles the numbers of sides and replaces the triangles with hexagons? Hexagons are a bit closer to the circle. By doubling the number of sides each time, the shapes inside and outside the circle would get closer and closer to the circle. In fact, mathematicians sometimes say that a circle is a regular polygon with an infinite number of sides. Archimedes didn't go as far as infinity. He stopped when he got to a shape with 96 sides. Using this shape, he was able to estimate that π lay between 223/71 and 22/7. This is where we get the approximation that most engineers use of 22/7.

As hard as they tried, no one was able to capture this enigmatic number completely and exactly with a fraction. Any fraction seemed to be either a bit too big or a bit too small. The reason is that π can't be written as a fraction. It isn't a ratio of two whole numbers. It is what we call an 'irrational number'. This has important implications for the challenge of physically realising a perfect circle. Quantum physics posits that space is not continuous but is made up of indivisible quantum units, a bit like pixels on a computer screen. These units are incredibly small. But if the universe is pixelated, or what scientists call quantised, any circle made out of these quantum units would imply that π can be written as a fraction. It means that, like the stone circles of Neolithic times, any attempt to create a perfect circle in reality is doomed to failure.

The idea of changing a circle into a square as Ahmes had done developed into a challenge that fascinated the ancient Greeks. Starting with a circle, could the tools of straight edge and compass be used to create a square whose area was exactly that of the circle? Given that the phrase 'to square the circle' has entered modern parlance as an expression of impossibility, you might guess that the Greeks failed in their attempts. But it wasn't until the late nineteenth century that it was finally confirmed that such a feat was indeed impossible. π is a 'transcendental number': in addition to its failure to be represented as a fraction, it is not even the solution to a simple algebraic equation. A straight edge and compass produces shapes whose dimensions are all solutions to algebraic equations. The transcendental nature of π means making a square whose area is π is beyond these simple tools. The simplicity of the circle hides inside it the complexity of this enigmatic number, π.

Painting the circle

If you have a compass, then drawing a circle is easy. But what if you don't have such a tool to hand? At the beginning of the fourteenth century, Pope Benedict XI was on the lookout for an artist who could paint something suitably impressive on the walls of St Peter's Basilica in the Vatican. He sent courtiers out to the great artists of the time to gather examples of their art to help him to choose whom to commission. One courtier was dispatched to Florence to visit the celebrated painter Giotto.

Giotto was recognised as one of the revolutionary artists of his age. He had created a new style of painting that moved away from the Byzantine style of previous generations. As the Renaissance artist and writer Giorgio Vasari later wrote, Giotto initiated 'the great art of painting as we know it today, introducing the technique of drawing accurately from life, which had been neglected for more than two hundred years'.

When the courtier arrived in Giotto's studio with his brief to bring back examples of this great new art to Pope Benedict XI, Giotto simply dipped his brush into some red paint and drew a perfect circle freehand onto a piece of paper. He passed the red circle to the courtier and told him to take it to the Pope as proof of his worthiness to paint the walls of the great basilica. As Vasari recorded:

> The messenger, seeing that he could get nothing else, departed ill-pleased . . . However, sending the other drawings to the Pope with the names of those who had made them, he sent also Giotto's, relating how he had made the circle without moving his arm and without compasses; the Pope . . . saw that Giotto must surpass greatly all the other painters of his time.

Giotto got the commission.

The power of the circle as a signature of a great artist had been used before in ancient Greece. Legend has it that Apelles of Kos had travelled to meet a fellow artist, Protogenes, on the island of Rhodes. When he arrived he found the studio of his friend empty except for an old woman. Rather than leaving a note to let him know he had come by, he is reputed to have painted a perfect circle on a blank canvas in the studio. When the woman asked who she should say had visited, Apelles simply replied, 'Say this came from me.'

When Protogenes returned, he saw the very fine line that the visitor had painted and realised that the circle was Apelles' way of letting him know he'd dropped by. To indicate his acknowledgement of the greeting, Protogenes added his own circle inside the first, with a line that was even finer than that of Apelles. Apelles made a second unsuccessful attempt to meet Protogenes, but when he saw the two concentric circles he added a third which perfectly filled the gap between the two. The battle of the circles!

It might be in homage to these two stories of Giotto and Apelles that Rembrandt included two perfect circles in the background of a self-portrait painted in 1665. Perhaps it was his way to subtly indicate his belief in his own prowess as one of the great painters of all time.

The three-dimensional circle

If drawing a circle was the mark of a great artist, then, for the architect, constructing a sphere was the ultimate test. A sphere can be thought of as a circle in three dimensions. Euclid had defined a circle as a curve whose points are all a fixed distance from a central point. The sphere is therefore a surface whose points are all a fixed distance from a central point in space.

The sphere, this three-dimensional circle, had both cosmic and earthly significance for our species. Aristotle was one of the first to realise that the planet we inhabited was the shape of a sphere. By observing lunar eclipses, Aristotle concluded that the shadows that the earth was making could most naturally be explained by the planet being spherical. Aristotle's sphere would usurp the Mesopotamian model of the earth as a flat circular disc. The spherical proposal was further confirmed by the observation that, when ships disappeared over the horizon, the hull would vanish first before the masts. One hundred years after Aristotle's proposal, the Greek mathematician Eratosthenes used the shadow cast by a stick at midday to calculate the circumference of this sphere with surprising accuracy.

The explanation for the spherical shape of the earth can be understood from observing smaller scale spheres that we see around us. If you blow a bubble, then although the shape might start out as some wobbling asymmetrical blob, the bubble eventually settles into the perfect shape of the sphere. Similarly, a drop of rain as it falls is not actually the tear shape that an artist might draw to depict its downward motion, but rather it forms a perfect sphere. Nature always tries to minimise energy, so in both cases it tries to find the shape using the least energy. The energy stored in the bubble or raindrop is proportional to the shape's surface area. Mathematics has proved that the sphere is the shape that has the smallest surface area needed to enclose a fixed volume of air or liquid, which is why the bubble and raindrop both aspire to be spherical. The formation of the earth has a similar explanation, although its spinning has resulted in the sphere being squashed at the North and South Poles and bulging out at the equator. The diameter of the earth measured from pole to pole is about 12,714 kilometres, and at the equator is about 12,756 kilometres, so the deviation is quite small.

Manufacturers of ball bearings and shot for guns have exploited this love nature has for the sphere. Ball bearings need to be as close to spherical as possible or else they will add friction to the system they are part of. Shot also needs to be as perfectly symmetrical as possible to ensure the flight is true when it leaves the gun. To make a perfect spherical metal ball, liquid metal is dropped from a great height. As it falls, the liquid tries to achieve the spherical shape that uses least energy. At the end of its drop it falls into a vat of cold water which solidifies the metal into the perfect sphere it achieved on its way down.

The ancient Greeks thought that spheres were also key to explaining the structure of the cosmos. The stars in the night sky were believed to be embedded in a huge celestial sphere whose surface was made from a transparent element called 'aether'. The fact that the stars did not seem to move relative to each other made this a very attractive explanation. The sphere itself would of course rotate, which explained how the stars moved across the night sky. It was proposed that the planets, which were observed to move relative to the stars, each had their own sphere which would revolve independently. The word planet has its origins in the ancient Greek for 'wanderer'. The model of the cosmos was this collection of concentric spheres.

As astronomical observations revealed the slightly unusual movement of the planets, corrective smaller spheres were added to take these anomalies into account. Aristotle's model, for example, had a total of 55 interconnected spheres in order to explain the paths of the planets. Copernicus's revolutionary suggestion made in 1543 to put the sun, rather than the earth, at the centre of the solar system helped explain the strange behaviour of the planets without the need for all these corrective spheres. Copernicus's thesis *On the Revolutions of the Celestial Spheres* changed our perspective on the cosmos, but it still maintained the centrality of the sphere to the fabric of the universe.

Hemispheres

It was the belief that the universe was a sphere which inspired the domes of Renaissance churches, such as Michelangelo's for St Peter's in the Vatican, completed in 1590, and Filippo Brunelleschi's for Florence's Duomo. When it was finished in 1436 the dome of Florence Cathedral was the greatest ever to be constructed from bricks and mortar.

The composer Guillaume Du Fay was asked to write a motet to celebrate the consecration of the cathedral on 25 March 1436. It was discovered in 1973 that the piece he wrote, *Nuper Rosarum Flores*, used the dimensions of the cathedral as a blueprint for the composition. There are four sections, each comprising two sets of 28 units made of four lots of seven-beat phrases. The four sections repeat the same music, but at different speeds, in a proportion of 6 to 4 to 2 to 3. All these numbers that Du Fay used to create the piece are the same numbers that define the proportions of the cathedral that Brunelleschi had worked on. A beautiful example of how the same numbers can be blueprints both for architecture and music.

When the commission to construct the cathedral's dome was given to Brunelleschi, many people were rather surprised. He was known primarily for his work as a goldsmith, not as an architect. Frankly, the architectural feat is still something of a mystery. Despite all the stunning architectural achievements of our modern era that populate the urban landscape, I'm still blown away every time I visit the Duomo in Florence.

Standing beneath Brunelleschi's dome certainly does give you the sense that you are at the centre of the cosmos. But there is a mathematical problem with this dream of building part of a perfect sphere because a hemispherical dome is not a self-supporting structure. An arch built in the shape of a half circle is doomed to collapse in the middle. The weight is too much for the supporting stones at the side.

But get the shape of the arch right and, as if by magic, the weight of the arch seems to be supported by the empty space beneath. What is happening in practice is that the load of the stones in the arch is being transferred to the abutments at its feet. Roman architecture is full of self-supporting arches but it took until the seventeenth century for mathematicians to identify the blueprint that was key to this architectural magic. The trouble was that it's not part of a circle.

In 1675, the English mathematician Robert Hooke published the 'true mathematical and mechanical form' of the ideal arch. Rather bizarrely though, Hooke wrote the theory as an anagram: *abcccddeeeeefggiiiiiiiillmmmmnnnnnooprrsssttttttuuuuuuuux*.

There were frequent controversies at the time over who should receive credit for certain scientific discoveries, including squabbles between Hooke and Isaac Newton. It's not quite clear why hiding his discovery in an anagram helped, but Hooke did this on a number of occasions. Perhaps it was to keep his findings to himself while still being able to claim priority once the cat was out of the bag.

It wasn't until after Hooke's death that his executor provided the unencrypted solution to the anagram: *Ut pendet continuum flexile, sic stabit contiguum rigidum inversum* – which translates to: 'As hangs a flexible cable so, inverted, stand the touching pieces of an arch.' In other words, the ideal shape of an arch is exactly that of a freely hanging rope or chain, only downside up. The equation for this curve was finally established by Gottfried Leibniz, Christiaan Huygens and Johann Bernoulli in 1691.

Although the mathematical proof of the power of the 'catenary curve', derived from the Latin for chain, took till the end of the seventeenth century, it was understood by those building arches and domes for many years before that. It is thought that Brunelleschi might well have used the principle in the construction of his dome in Florence.

A striking fact about Brunelleschi's dome is that the space it spans is so vast there wasn't enough wood available to build the scaffolding that would normally be required to support it during construction. Instead Brunelleschi had to come up with a design that would be self-supporting not only once it was completed but also as it was being built.

One of the tricks that Brunelleschi used was to build two domes not one. The same principle was employed by Christopher Wren when he built St Paul's Cathedral, but Wren went one further and used three domes. The external dome of St Paul's is a hemisphere representing the cosmos. The internal dome exploits the catenary curve and creates an interesting illusion of depth for those inside the cathedral. But the work of holding up the whole structure is done by a third, conical dome, hidden from view, which sits between these two aesthetic domes.

Brunelleschi was very secretive about his methods, but an analysis of the building in 2020, published in the journal *Engineering Structures*, has shed some light on the innovative ideas Brunelleschi came up with. To construct the brick walls of the dome, Brunelleschi employed a novel herringbone pattern that allowed the bricks to self-reinforce as they were being laid, so that they wouldn't fall off the wall as it became more inclined. The amazing thing is that Brunelleschi's clever idea developed in the fifteenth century could help architects of today who are interested in using drones to build structures rather than wasting resources building expensive scaffolding. It could lead to much more sustainable architectural design and construction methods, all thanks to the ideas of a Florentine goldsmith who turned his hand to building domes.

The power of the catenary curve and these upturned chains as a blueprint for architects has continued to this day. Antoni Gaudí employed the technique extensively in the construction of his famous Sagrada Família. You can see the blueprint of these hanging chains implemented in a fascinating model in the cathedral's museum. If you attach a weight to a hanging chain then this alters the shape of the curve, pulling it down. But this in turn determines the shape of the arch that would support the same weight placed on top of it. Using this system of ropes and weights, Gaudí could play around with the possible shapes that nested vaults might assume.

It's not just the mathematics of arches that Gaudí enjoyed playing with in the Sagrada Família and beyond. The streets of Barcelona are paved with beautiful hexagonal tiles with fossil-like imprints that Gaudí designed. The repeated use of spirals in the tiles and throughout his iconic Casa Batlló reflects the movement of water and air in nature. 'Originality is returning to the origin. Nature is my master,' he once declared. One of the entrances to the cathedral boasts a 4×4 magic square, where the rows, columns and diagonals contain numbers which add up to 33, the age at which Christ is meant to have died.

Magic squares crop up in a number of artistic settings; one of the earliest is in Dürer's engraving *Melencolia I*. It is another 4×4 square, with the numbers 1 to 16 arranged to add up to 34. The numbers in the bottom row of the square include 15 and 14, to mark the date it was created: 1514. A different size of magic square was associated with each planet and the qualities it represented. Dürer used the square signifying Jupiter, which represented optimism, hope and growth, perhaps to counter the melancholy depicted in the rest of the engraving.

Building spheres

Building the hemispherical domes of the Renaissance certainly pushed the architects of the time to their technological limits, but the ultimate challenge for an architect is to complete the shape and construct a whole sphere. This was the dream of architects in Paris at the end of the eighteenth century. In 1784, Étienne-Louis Boullée had drawn up plans for the construction of a huge sphere in Paris dedicated to Isaac Newton. Boullée's proposal was for a hollow sphere 150 metres high, making it taller than the Great Pyramid of Giza. The architectural drawings that he made for it are stunning but also illustrate how crazy the idea of building a sphere would have been at the time. The whole thing looks totally surreal. The surface would be peppered with small holes to allow in light during the day. The aim was to create a picture of the night sky inside the sphere, rather like a planetarium. At night the interior would be lit by a huge lamp suspended in the middle of the sphere, like the sun that sat at the centre of the eighteenth-century universe.

It is perhaps curious that a French man would want to honour an English mathematician in this way, but Boullée had chosen to commemorate Newton in this cenotaph because he represented a beacon of the Enlightenment. His idea was that an empty sarcophagus for Newton would be constructed on the floor of the sphere. Boullée was so enamoured by the project that he produced many ink-and-wash drawings which were engraved and circulated widely.

For Boullée, the sphere was the most beautiful and natural shape. For the revolutionaries who came to power in France a few years after his proposal, the shape perfectly encapsulated the spirit of the revolution. *Egalité*. No direction favoured over any other.

It took another 200 years for a sphere eventually to be built in Paris, in the Parc de la Villette. Even today, the architectural feat has a sense of the unreal about it. I remember the shock I felt when I first saw this huge silver sphere, which looks like an alien spaceship that has just landed. The eighteenth-century idealists, however, might be disappointed to discover that the sphere was actually built to house an IMAX cinema. Not a terribly revolutionary venture. Called La Géode, this sphere is made up of 6,433 triangular pieces to create the impression of curvature.

The underlying shape that is the blueprint for this sphere is one that we will meet later in the book: the icosahedron, made up of 20 triangular faces. By subdividing these large triangular faces into ever smaller triangles, an architect can give the illusion of a sphere. Nature plays a similar trick when building viruses. The rotavirus that causes bad diarrhoea is made up of 120 triangular pieces made of proteins, arranged into a spherical shape. The sphere in nature is a low-energy shape but also has protective strength, which is why eggs are almost spherical.

The triangle is also a strong mathematical shape from which to build these spherical structures. It has a rigidity that makes it robust, unlike the square, which has a tendency to deform into a parallelogram. This is why in the geodesic domes popularised by Buckminster Fuller you see a framework built out of multiple triangles.

The spheres and hemispheres of the architectural realm are representations on earth of the spheres that the ancient world believed were the blueprints of the universe. But the ancients also believed there was an underlying harmony to that blueprint which gave rise to the concept of the music of the spheres. So it is perhaps not surprising to find circles at work in the music that we have composed.

Musical loops

The temporal element of music means that a piece has a beginning, middle and end. There is a progression that mirrors the linearity of literary narrative. A piece often starts by introducing themes, then develops them before ending with a climactic finale. But composers have also explored the interesting possibilities of loops in music.

Johann Sebastian Bach is often heralded as the composer who more than any other enjoyed mathematics as a blueprint for composition. Lots of interesting numerical games run through his work. He enjoyed using simple algorithms to generate complex outcomes. Indeed, some of his pieces were written in code that a player is meant to unravel in order to perform the piece.

For example, the keyboard work known as the 'crab canon' from Bach's *The Musical Offering* is written as a single line of music. But when you get to the end, you see that Bach has repeated the clef and key signature that began the piece but has written them in reflection. Bach is instructing you to play the piano piece from the beginning with one hand and play it in reverse from the end with your other hand. When both parts are played together, the piece makes a beautiful two-part invention, all made out of one line of music.

The Musical Offering was inspired by a challenge set by Frederick the Great for Bach to improvise a fugue on a theme that the king tapped out on his new fortepiano. Bach's reply was an extraordinary fugue based on six voices. But subsequently he also composed ten canons that have a very algorithmic mathematical character to them. He provided musical seeds and mathematical rules for growing these seeds.

There is one work that Bach wrote where the idea of a loop is introduced as a blueprint for the composition. The Goldberg Variations is a piece for keyboard named after the harpsichordist for whom it was written. Count Kaiserling, the Russian ambassador to the court of Saxony, would often travel with his harpsichordist Johann Gottlieb Goldberg. The count suffered terrible insomnia and, rather than ruminating in silence, he would employ Goldberg to spend the night playing in an antechamber.

On a visit to Leipzig, home to the great Bach, the count was keen for Goldberg to receive some instruction from the composer. He mentioned that he would love some new pieces for Goldberg to play which, the German musicologist Nikolaus Forkel reported, 'should be of such a smooth and somewhat lively character that he might be a little cheered up by them in his sleepless nights'. The result was a set of 30 variations based on the aria that opened the piece.

But rather than ending on the 30th variation, Bach joined the whole piece up in a circle by getting the performer to finish by playing the aria again. It's almost as if the musician at the keyboard was being invited to begin the journey again – perhaps important if the count had still not fallen asleep. Bach further muddied the sense of where the beginning of the piece is by naming the 16th variation an overture. This variation is halfway through the cycle, yet an overture is normally a piece which begins a musical composition. The naming of this central variation as an overture stresses even further the idea that perhaps one doesn't know where the beginning or ending of this piece is. Like a circle, there isn't a starting point.

Together with this primary loop Bach introduced another, more subtle circle into the composition. Every third variation is a canon. School music lessons often feature canons where one half of the class starts singing a song like 'London's Burning' and then, after a few bars, the second half of the class starts singing the same song from the beginning but delayed in time. This is a very familiar structure in music and one that Bach enjoyed playing with. In a musical score, it looks a little like a musical tile repeated to the right of the original tile.

The third of the Goldberg Variations is a classic example of a canon where the performer begins to play a tune and, after a bar, you hear a second voice play the opening of the tune again against the development of the first voice. But when Bach got to the next canon, the sixth variation in the piece, he decided to do something slightly different. This time, when the second voice comes in, instead of simply repeating the first voice, it starts one note higher in the scale. It's as if the musical tile has shifted both right on the score and up one step. The geometry is changing.

As the piece progresses, so the second voice climbs through the canons. Each new canon sees the second voice starting higher and higher in the scale. But then something rather magical happens when we reach the eighth canon, variation number 24. The musical scale that Bach was using to climb through the piece has seven notes and then, on the 8th note, it returns to the starting note but an octave apart. The notes sound so similar that we give them the same name.

But the effect of these climbing canons in the Goldberg Variations is that once we arrive at this eighth canon, it's as if we've come round full circle to where we started. It's like that example of Escher's staircase, where the hooded figures think they are climbing or descending a staircase, only to find that once they've completed four sides of the building they are back to where they started. This is an example of what Douglas Hofstadter, in his famous book *Gödel, Escher, Bach*, named a 'strange loop': a loop that seems to be ascending through layers of some hierarchy, only for the climb to bring you back to the bottom layer where you started. It's not surprising that Hofstadter chose Bach as his musical representative for his exploration of strange loops.

I must admit that when I first listened to the Goldberg Variations, as beautiful as the piece is, I would always get rather lost inside this quite complex extended set of variations. But once I understood how these circles worked to create the underlying blueprint for the piece, it really helped me to listen in a more engaged way. The blueprint provided a helpful set of signposts that allowed me to navigate the twists and turns of the double-looped structure as one circle runs through the 32 movements from the aria to the central overture and back to the aria again, while the second circle runs through the notes of the scale in the canons in order to return to the octave.

Yet these independent loops in the Goldberg Variations also hint at an interesting new blueprint hiding in Bach's great keyboard piece. The torus.

The torus: a circle's worth of circles

If you take a circle and spin it around its diameter, then it maps out a sphere. But if you take a line outside the circle and spin the circle in three-dimensional space around it, you get a different three-dimensional shape: the torus. This bagel shape is defined sometimes as a circle's worth of circles. It is one of my all-time favourite mathematical objects.

The torus has many interesting properties with deep connections to number theory, geometry and topology. The most significant feature is that there are two loops in this structure which don't overlap. If I take any point on the surface of a torus, then I can head in a loop that circles around the shape. Or I can take a different direction and head in a loop through the hole in the middle and round the other side. These two loops do not meet each other on their respective journeys. This is a defining feature of a shape that has a hole in it, like the torus. A sphere is a shape with no hole; if you try to head off in circles around it from a point on the surface, then you find that the paths always meet at the point opposite your starting position.

These independent loops are why I often feel like the torus is the blueprint that underpins the Goldberg Variations, rather than, say, a sphere. A torus also turns out to be the shape at the heart of one of Borges's most famous short stories, 'The Library of Babel'. The story opens with a description of the shape of the library: 'The universe (which others call the Library) is composed of an indefinite and perhaps infinite number of hexagonal galleries . . . From any of the hexagons one can see, interminably, the upper and lower floors.' The layout of the library exploits the hexagonal blueprint used by the honey bee to make its home. But it's the blueprint that describes how this beehive is laced together that taps into the shape of the torus. As the opening of the story declares, the library is a metaphor for our universe, and Borges uses a literary narrative of ten pages to come up with a possible solution to a question that has challenged scientists for centuries: what is the shape of our universe?

Many people are rather flummoxed by this question. How can the universe have a shape? Isn't it infinite – doesn't it just go on for ever? But for those who think that the universe can't be infinite, there is a real challenge. How does it work? As I explained earlier, the ancient Greeks tried to solve this by encasing the universe in a huge ball in which the stars were embedded. But that only raises the secondary problem of what's beyond the wall. We're not living in *The Truman Show*, with some celestial film crew beyond the edge of the universe watching our every move.

As Borges's narrator explores the library, he begins to realise that it does not go on for ever. He believes the books between them contain every possible combination of letters in that alphabet, and therefore that 'the Library contained all books'. Each book in the library is 410 pages long, but he also thinks that each book is unique. Although there are a lot of books in Borges's library, more than the number of atoms in our observable universe, there are still only a finite number of books. But then if the library is finite, what lies beyond? Just as Borges had learned from reading his books of modern science, a finite universe is destined to repeat itself.

The narrator's first idea is some version of a sphere: 'The Library is a sphere whose exact centre is any one of its hexagons and whose circumference is inaccessible.' Borges loved talking about the idea of spheres with no centres, a notion which goes back to quotations of philosophers such as Pascal: 'Nature is an infinite sphere whose centre is everywhere, whose circumference is nowhere.' But ultimately the librarian comes up with a different solution:

> I venture to suggest this solution to this ancient problem. *The Library is unlimited and cyclical.* If an eternal traveller were to cross it in any direction, after centuries he would see that the same volumes were repeated in the same disorder.

These cycles are not around a sphere, where they would intersect each other on their journey, but around a torus, where they head in independent directions that don't intersect. What is extraordinary is that this idea of wrapping up a universe so that it is cyclical is the same solution that mathematicians in the twentieth century came up with for answering the challenge of how our universe might be finite but without walls.

To understand Borges's solution it is instructive to take a toy model of a universe. The video game *Asteroids*, which was created by Atari in 1979, serves as a perfect example of a two-dimensional universe that is finite yet without boundary. The universe just consists of your computer screen, but when a spaceship heads towards the top it doesn't bounce off the edge, like a two-dimensional *Truman Show*, it seamlessly reappears at the bottom. As far as the astronaut inside the spaceship is concerned, they are just travelling endlessly through space. The same rule applies if the spaceship heads to the left-hand side of the screen. It doesn't encounter a wall but simply reappears on the right.

The exciting revelation is that this *Asteroids* universe has a recognisable shape. If I allow myself a third dimension within which to wrap up the universe of *Asteroids*, I can join the top and bottom of the screen to make a cylinder. Then since the left- and right-hand sides of the screen are also connected, I can join the two ends of the cylinder to create a torus. The surface of this three-dimensional shape is the finite universe in the game of *Asteroids*. So, in this way, something can be finite but cyclical.

Borges, though, goes one better: he has conceived the Library of

Babel to be like a three-dimensional game of *Asteroids*. His solution to the shape of the library is that it is actually the surface of a four-dimensional bagel, or what we call a 'hypertorus'. It's beautiful. A literary version of a shape we'll never see. So how did Borges come up with such an extraordinary idea? We know that he had books of mathematics in his own library in Buenos Aires which tried to give readers a sense of the fourth dimension, but he was somewhat confounded by the mathematical language that described the shapes that exist in this four-dimensional space. So instead he used his own language of story-telling to navigate these geometries of the mind.

The fourth dimension became quite an obsession for Borges. For him, these shapes were liberating. Given that he was blind by the time he was writing many of his later stories that reference the fourth dimension, it is very striking to conjecture whether his no longer being able to see shapes in three dimensions allowed him a freedom to move into other worlds that were unseen.

In an essay entitled 'The Fourth Dimension' he explains how a point that is imprisoned in a circle in two dimensions can nonetheless escape by exploiting the third dimension of height without touching the circle. By the same token he argues that a man trapped in a dungeon could use the 'unimaginable' fourth dimension to break out without touching the walls, floor or ceiling. 'To deny the fourth dimension is to limit the world, to affirm it is to enrich it.'

The protagonist in Borges's short story 'There Are More Things', published in 1975, talks of his attempts with his uncle to understand the work of Charles Howard Hinton, the scientist above all others who wrote about the fourth dimension for the general public and from whom it seems clear Borges learnt most about these ideas.

Hinton had written a seminal book at the beginning of the twentieth century entitled *The Fourth Dimension*, which his publisher subtitled *Ghosts Explained*. Hinton believed that this fourth dimension might actually exist at very small scale in the brain and perhaps could even allow a way for different consciousnesses to connect. His text came with beautifully illustrated colour cubes which were somehow meant to give one a view of the hypercube living out in the fourth dimension. Hinton's blossoming career as a public intellectual, however, came crashing to the ground when he was convicted of bigamy and sent to jail. Although the existence of a fourth dimension mathematically provides a way to escape a three-dimensional cube, Hinton's exploration of hyperspace didn't extend to him finding a way out of his three-dimensional cell. Despite this personal downfall, his writings caught the public imagination at the time.

Borges explicitly referred to them in 'There Are More Things'. The story's narrator recalls how his uncle had lent him Hinton's treatise that attempts to prove that space really has a fourth dimension. Together they use Hinton's coloured cubes to try to understand this hidden dimension. 'I shall never forget the prisms and pyramids we erected on the floor of his study.'

The title of the story is a reference to Hamlet's assertion that 'There are more things in heaven and earth, Horatio,/Than are dreamt of in your philosophy'.

It was published as part of a collection that Borges titled *The Book of Sand*. The story that gave its name to the collection opens with another way to understand the idea of dimensions beyond our three-dimensional physical universe.

The narrator recognises that you get one dimension by putting together infinitely many points of zero dimension. Then, by piecing together infinitely many lines, you get a second new dimension. This new, two-dimensional plane can then be stacked to get a third dimension. But then why can't one just continue this whole process and stack three-dimensional volumes in order to extend space into a fourth dimension?

Having started the story with this highly mathematical burst explaining how to create the fourth dimension, the narrator decides that this geometric opening 'is definitely not the best way to begin my tale. Affirming a fantastic tale's truth is now a story-telling convention; mine, though, is true.'

This idea of how you go from line to plane to volume, or from point to circle to sphere, is at the heart of how mathematicians have created a language to talk about hyperspace, or the idea of a hypersphere living in four dimensions. I will explore this idea in more detail in a later blueprint, when we consider how to make Platonic solids in hyperspace, but here is a taster of how a mathematician makes a hypersphere in four dimensions.

The equation for a circle is the set of points in the plane whose coordinates (x,y) satisfy the equation:

$$x^2+y^2=1.$$

The formula on the left of the equation represents the square of the distance from the origin, the central point of the plane. The number on the right asserts that all these points are the same fixed distance from this point. Hence they form the circle that Euclid first defined. In three dimensions, we have a third coordinate, which we can call z, representing height above the plane in which we drew the circle. Distance from the central origin in this three-dimensional space is captured by the same equation now in three variables:

$$x^2+y^2+z^2=1.$$

If the distance of all the points is again fixed, as this equation asserts, then the points all sit on the surface of a sphere of radius 1.

But just as Borges did, the pattern of how to push into the unseen fourth dimension has been established. Add a new variable, w, which measures how far one has travelled in this new unseen direction. The distance of points in the four-dimensional space is determined by the same equation of squares. Hence if we want a hypersphere – a shape in four dimensions where all the points are a fixed distance from the central origin – its equation will be:

$$x^2+y^2+z^2+w^2=1.$$

For Borges, these equations left him cold. Rather it was the narrative of the library that allowed him to explore these shapes in hyperspace.

Twisted circles

Before we leave the blueprint of the circle there is one more twist in the story. A literal twist. Borges referred to a 'mildly obscene' shape that he encountered in his mathematical reading called a Möbius strip. This is a loop with a twist in it. If you take a long strip of paper and you twist the paper before gluing both ends together to make a loop, you get the shape called a Möbius strip, named after the German mathematician August Möbius who discovered the shape in 1858.

The twist causes the Möbius strip to possess some rather crazy features. For example, if you start to colour one side of the shape, you discover when you come back to where you started that the whole shape has been coloured. It only has one side, unlike a simple untwisted loop, which has an inside and an outside. Take a pair of scissors and cut down a line running round the strip. Instead of the two loops of paper that you'd expect by cutting something in half, the result is a single piece of looped paper with two twists in it. Make the Möbius strip from clear plastic and place a cut-out of a right hand on it. Now move the hand round the strip. When it returns to the position where it started, the hand has flipped over to become a left hand. This is why we call the surface 'non-orientable'. It flips things over.

This idea of the shape flipping things over has made the Möbius strip an attractive blueprint for a number of writers. *The Bald Soprano*, a play by Eugène Ionesco, and *Moebius*, a film by Argentinian director Gustavo Mosquera, both explore the idea of how to use the interesting shape to tell a story. Premiered in 1996, *Moebius* is a classic science fiction movie, where a subway train is lost because it turns out the tracks have been laid in the shape of a Möbius strip. It is based on the short story 'A Subway Called Moebius' by the scientist A. J. Deutsch, published in 1950, the same year as the first performance of Ionesco's play. In the film, the Möbius strip acts as a political metaphor for the people who went missing in the 1970s and 1980s during Argentina's military dictatorship.

A Möbius strip made out of train tracks can actually be experienced if you go to the funfair in Blackpool. When you get in the cars at the beginning of this rollercoaster ride it looks like there are two tracks running parallel to each other. The ride is named after the Grand National horse race and is meant to create the sensation that your car is racing against the car on the adjacent track. But the curious feature of this rollercoaster is that, when you return to the beginning of the ride, your car is now on the other track.

What has happened is that the tracks have swapped over, so in reality there is just one long line of track. Think of a strip of paper running between the tracks. At some point, a twist has been put in this piece of paper as the ride plays out, such that when it joins up back at the beginning, the shape is a Möbius strip. Just as a Möbius strip has only one side, it also has just one edge, which in the case of the rollercoaster is the single train track that the cars race along. Built in 1936, this is one of just two existing Möbius rollercoasters on which you can experience this twisted loop.

While a literal Möbius strip acts as a plot device in Mosquera's movie, the way Ionesco uses the idea in *The Bald Soprano* is perhaps more interesting, because the Möbius strip is used to shape the play. The play doesn't end but returns to the beginning, although we discover that the two couples in the play have swapped roles, as if they have travelled along a Möbius strip.

The same idea is at the heart of a play that I wrote with actress Victoria Gould, called *I Is a Strange Loop*. We encountered the idea of a strange loop in Bach's musical staircase that seems to be climbing yet returns to its beginning again. Taking its inspiration from Borges's 'The Library of Babel', the play is set inside a cube in which the character X lives. For X, this cube is his universe. He is unaware that there is anything beyond the walls of his cube until a second character, Y, bursts through a hidden door. It turns out that X's cube is just one of a series of cubes that are connected in a long corridor. Y is desperately looking for a way out, a final cube beyond which she hopes there is something else.

Y tries to persuade X to join the search for 'out', but he is convinced that the line of cubes is infinite and Y will be journeying for ever. Y leaves X behind, determined to prove X wrong. But it turns out that they are both wrong. Having exited stage left, Y re-enters after some time through the door stage right. The universe that they inhabit is 'unlimited but finite'. It is a loop.

As the play progresses, they explore different directions that might provide a way out, only to find their universe looped in all directions. The play has the shape of a four-dimensional torus. But as the two characters interact, they start to hanker after the life of the other. X wants a world of sensation and embodiment and, ultimately, mortality. Y wants the abstract mathematical world that X inhabits, with its promise of infinity. They end up creating a Möbius script which allows them to exchange places. As they navigate the loops in the play, they ultimately discover that some of the loops are twisted into a Möbius strip, swapping things over. Left becomes right. X becomes Y.

We saw how Bach loved the idea of circles in music, as he realised in the Goldberg Variations, his masterpiece for keyboard. In 1974, 14 new canons by Bach were discovered attached to a copy of the first edition of the Goldberg Variations in a private collection in Paris. The music was written in Bach's own hand and headed 'Various canons on the first eight bass notes of the preceding aria'. The canons were not, however, written out in full, but were offered as coded challenges to the performer. They were annotated to indicate how to grow the seeds into fully flowered pieces.

It has been speculated that Bach wrote these puzzle canons to submit to the Corresponding Society of Musical Sciences upon his admission as a member in June 1747. The society was a mix of musicians and scientists, and the puzzles would have appealed to the mathematically minded members. The portrait that Elias Gottlob Haussmann painted of Bach to mark his admission to the society has the composer holding a copy of the 13th of these canons in its coded form.

But among the 14 puzzle canons that he wrote to accompany the Goldberg Variations there is one very curious example. Because when you unravel the code of the fifth canon, you discover it is a musical Möbius strip. The pianist is expected to play the following:

The bottom line looks quite different to the top until we do something rather strange. Let's peel off the top line. Imagine the top notes are written on a transparent piece of celluloid which is then twisted into a Möbius strip. When we go round the strip, the first half produces the first line; but as we are taken onto the back of the celluloid and read the notes from the reverse, we get the second line of music. Here is a genuine use of a Möbius strip to make a canon. Bach had been drawn aesthetically to this mathematical blueprint more than a century before the mathematician Möbius articulated its structure.

Our next blueprint turns out to be another structure that was discovered by musicians long before mathematicians realised its importance.

Blueprint Three

The Fibonacci Numbers

Madame Gambu-Moreau ran up to the architect, not a trivial feat given that she was heavily pregnant. She was about to become one of the first inhabitants of the new apartment building that was being opened that day with a big ceremony. 'Sir, would you do me the honour of being the godfather to my child?' The baby was to be the first child born within the building. Since the architect believed that his building was meant to encompass all aspects of life, he agreed. That baby, Gisèle Gambu-Moreau, went on to grow up in the building and have her own daughter, who moved into one of the other flats and in turn had her own child: the fourth generation of the same family to call the building home.

Like the Gambu-Moreaus, who moved into L'Unité d'Habitation on 14 October 1952, I am a big fan of this building. But when I show people a picture of it, they generally recoil in disgust at this huge concrete tower block situated on the edge of Marseilles. For many, it represents everything that went wrong with architecture in the 1950s and 1960s. The term 'brutalist' was popularised by the British critic Reyner Banham to describe this modern style. Many people mistakenly believe that it is meant to express the horror most feel towards this fad for huge concrete edifices. In fact, the term is a pun on the French *béton brut*, meaning raw or rough-cast concrete. This was the medium of choice for many of the buildings that the architect of L'Unité d'Habitation would construct. The architect's name: Le Corbusier.

Although in fact this wasn't his name. He was born Charles-Edouard Jeanneret, but he decided to reinvent himself as he emerged as an architect in the 1920s by giving himself the name 'Le Corbusier'. The origin of the name is unclear. There is a suggestion it was the name of one of his ancestors, although others have enjoyed the idea that it derives from *courber*, the French for 'to bend'. Le Corbusier enjoyed bending buildings as much as bending others to his will. But this alter ego was meant more as a protective shield as the young architect began to promote his challenging views on architecture to the world.

These included a proposal in 1925 to demolish a large swathe of the centre of Paris on the right bank of the Seine and replace it with a network of 18 identical skyscrapers. Called the 'Plan Voisin', it was Le Corbusier's response to the huge surge in urban poverty that he witnessed in cities such as Paris, together with the explosion in the number of cars, which was beginning to snarl up the roads. His radical idea is often used to illustrate the idea of 'Le Corbusier the megalomaniac'. Destroy the elegant Place des Vosges, the grand Palais-Royal! People were outraged. But Le Corbusier never intended the destruction of such celebrated buildings, but rather their incorporation into a modern city that could accommodate the growing urban population.

Although his plan for modernising Paris never came to fruition, the building blocks at the heart of the plan would finally be realised in Marseilles. L'Unité d'Habitation was a concept that Le Corbusier believed could serve as a blueprint for many sites of urban living. Indeed, five other Unités have been built, although Le Corbusier refused to acknowledge the Unité built in Berlin because the planning authorities insisted on changing the dimensions so that they did not fit Le Corbusier's strict framework. He hadn't reckoned on the intransigence of German building regulations.

L'Unité d'Habitation – or La Cité Radieuse, 'the radiant city', as Le Corbusier preferred to call it – is a tower block with 17 floors plus a rooftop terrace. It was built to house 1,600 people on low incomes displaced by the bombing of the Second World War. The first thing that strikes you is that this huge edifice is on stilts, or what Le Corbusier called 'pilotis'.

These thick, tapering, reinforced concrete blocks give the impression of grey elephants carrying the building on their back. Pilotis were one of Le Corbusier's 'five points of architecture', which he first proposed in a manifesto written in the 1920s. The idea was to raise the lower floors of the building to allow air to circulate underneath them and prevent dampness from the ground seeping into them. But they also served an aesthetic function, providing unobstructed views through the building 'all the way to the horizon' as you approach. The space could also be turned into a garden, or used for parking the cars that were now part of city life.

The other striking quality of the building that greets you is the riot of colour across the façade: a vibrant and intense range of reds, blues and yellows. The colours run throughout the building. Each apartment has its own colour scheme, chosen from a carefully selected palette of 43 colours arranged into 14 series which Le Corbusier used for his buildings. 'Colour is a factor of our existence,' Le Corbusier believed; it inevitably triggered an emotional response in the viewer. There doesn't appear to be any clear pattern to the arrangement of colours on the side of the Unité building in Marseilles, which gives it a feeling of washing being hung out to dry in the Mediterranean breeze.

The way the building floats above the ground makes it look rather like a huge ocean liner, and that was also a vision that Le Corbusier had for the Unité. Just as an ocean liner must be self-sufficient, so too should the Unité. It should function as an independent, isolated city in its own right which the 1,600 residents need never leave. Le Corbusier even designed the ventilation stack on the roof terrace to resemble a huge concrete funnel to reinforce this idea of the building being a huge ship.

The Unité was built to combine the private and the public. Apartments would provide the private space, but there were also communal spaces to allow residents to come together. The terrace was part of this vision of shared public areas and included a garden that was also a running track, a theatre for concerts or dramatic performances, even a kindergarten raised on more pilotis so that it floats above the whole building. There was also a small paddling pool decorated with coloured tiles, not big enough to do lengths in but certainly a welcome place to cool off from the summer heat. The whole place feels like a wild concrete playground, with every corner filled with architectural games. The common areas continued within the building, where the seventh and eighth floors served as an internal street housing a hairdresser, a grocery, a butcher, a bakery, medical facilities, and a hotel for guests of the residents to stay in. This was a city within a city.

But it is the private living spaces that are most important for the residents' sense of well-being and here is where Le Corbusier's ideas truly excelled. Each flat is on two floors. The level one enters from the central corridor extends half the width of the building; one then ascends or descends to the second level, which runs the whole way across the block. The Unité was constructed along a north–south axis, so that the flats look out to the east and west, which gives them maximum benefit from the sun. In Marseilles, one side faces the mountains while the other faces the sea.

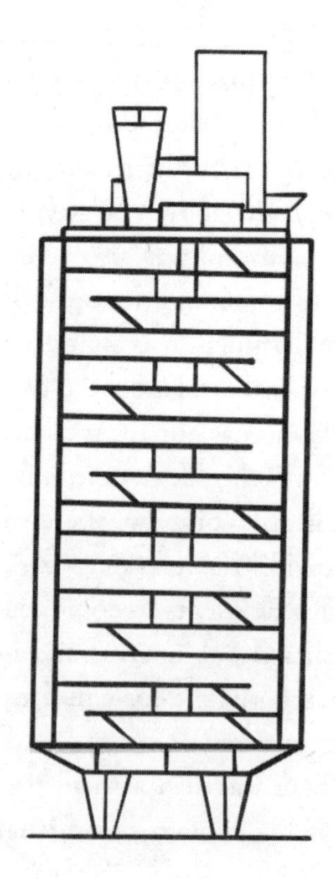

There is a clever asymmetry in the way that the flats are interlocked so that one corridor serves three floors meaning more space can be given to the flats. Each flat consists of a master bedroom, two single bedrooms and a bathroom on one level. On the other is a living space which extends the full height of the flat, and a kitchen which is kitted out like a laboratory. With the kitchen, Le Corbusier paid particular attention to ensuring that the space facilitated the domestic work of the family, still generally done by women in the 1950s.

So what is the big deal about this tower block? The secret is actually carved into the wall as you enter the building. Incised into the raw concrete is the image of a strange man in a hat with his arm in the air as if greeting visitors. Three other similar figures are stacked on top of him. Beside them is a strange shape carved into the wall whose edges consist of two wavy lines. The wavelengths of these oscillating lines increase as they ascend the building. This is Modulor Man and the wavelengths are the key to the blueprints for the flats.

The distances marked out by these waves measured in millimetres give rise to two sequences of numbers. The first sequence Le Corbusier called his red series:

6, 9, 15, 24, 39, 63, 102, 165, 267, 432, 698, 1,130, 1,829.

The second sequence was his blue series:

11, 18, 30, 48, 78, 126, 204, 330, 534, 863, 1,397, 2,260.

This final number, 2,260 millimetres, was actually the starting point for all of Le Corbusier's ideas. This was the height of the man with his hand in the air and was the measurement Le Corbusier used for the distance between the floor and the ceiling in the flats of the Unité (and double that in the living area). But that was just the beginning. All the measurements that ran through the building were determined by the numbers in these two series.

Apart from a few slight anomalies, there is a fundamental pattern that underlies these sequences of numbers, and it is one that is very familiar to mathematicians. It is the Fibonacci rule.

Nature's favourite numbers

1, 1, 2, 3, 5, 8, 13, 21, 34, 55, 89, . . .

The Fibonacci numbers. Each number is obtained by adding together the two previous numbers in the sequence. If you look back at Le Corbusier's series, you can see that the same rule applies. What is perhaps unsettling to those who have encountered the Fibonacci numbers before is that Le Corbusier's numbers begin with 6 and 9 rather than the more accustomed starting point of 1 and 1. But the rule is the same.

The Fibonacci numbers are, without a doubt, nature's favourite numbers. They crop up time and again across the natural world. The classic example is the number of petals you find on flowers. Invariably it is a Fibonacci number. Trillium flowers have 3 petals, a pansy has 5, a delphinium has 8, marigolds have 13, chicory has 21, pyrethrum 34, and sunflowers often have 55 or even 89 petals. Some plants have flowers with double a Fibonacci number of petals. There are plants, like some lilies, where the flower is essentially two flowers, one on top of the other. And if your flower doesn't have a Fibonacci number of petals, then that's because a petal has fallen off . . . which is how mathematicians get around exceptions. But seriously, there certainly are exceptions to this rule. Biology is never mathematically perfect. For example, there is a member of the primrose family called a starflower which has seven petals, but across the plant kingdom this is actually quite rare.

The recognition of a link between the petals of flowers and the Fibonacci sequence probably dates back to the early seventeenth century. The German astronomer and mathematician Johannes Kepler wrote about the connection in his 1611 essay *On the Six-Pointed Snowflake*. Although it was only remarkably recently that anyone was able to offer a clear explanation for why these numbers appear and others don't. It turns out to be related to my fourth blueprint, and I will save the explanation until then.

The Fibonacci numbers also appear quite frequently in the fruit that many plants bear. If you cut open a fruit, you often find these numbers underlying the way the fruit is segmented. Cut a banana in half and you'll see a three-pointed star. An apple cut horizontally will reveal a five-pointed star at its core. The bright-orange persimmon fruit has an eight-pointed star hiding inside. Other fruits wear their Fibonacci numbers on the outside. If you look at a pineapple, you will see that its skin segments follow three obvious spirals running at different angles across the fruit. These spirals often consist of consecutive Fibonacci numbers of segments – a fact first revealed to the world in the *Pineapple Quarterly* in 1933 (no joke, there really was a *Pineapple Quarterly*).

In smaller pineapples, the steepest spiral often has 5 segments, the next 8 and the shallowest spiral 13. But when the pineapple grows bigger, we find these increasing to an 8–13–21 sequence, which is encountered in the majority of pineapples. After reading the 1933 study, Philip B. Onderdonk of Philadelphia, Pennsylvania, went in search of a mega-pineapple with a 13–21–34 sequence in its segments, but as he reported in a paper published in the *Fibonacci Quarterly* in 1970: 'No such pineapples were ever found.'

These numbers seem to be the blueprints for many structures across the natural world. The reason they are named after Fibonacci, a thirteenth-century Italian mathematician, is because he discovered another place that nature uses them. They crop up in helping to understand population dynamics especially in relation to rabbits.

Leonardo Pisano, better known as Fibonacci, was the son of a customs' official and as a child he had travelled with his father around North Africa, where he learnt about the developments of Arabic mathematics and especially the benefits of the Hindu–Arabic system of numerals, the numbers we use today. When he got home to Italy, he wrote his *Liber Abaci*, or *Book of Calculation*, which is credited with kick-starting mathematics in Europe in the Middle Ages.

Fibonacci demonstrated how simple the new system was compared to the Roman numerals that were in use across Europe. Calculations were far easier, a fact that had huge consequences for anyone dealing with numbers – pretty much everyone from merchants to mathematicians. In the second part of his book, he presented a whole series of challenges, puzzles and exercises to stimulate the mathematical mind of the merchants the book was aimed at. And it is here that we come across the puzzle that gives rise to the numbers that are named after him. It's a puzzle about a growing population of rabbits.

You start in the first month with a juvenile pair of rabbits. It takes a month for them to mature, so in the second month there is still just this single pair of rabbits. But now they give birth to a second pair of rabbits. So in the third month there is a mature pair of rabbits and a new pair of juvenile rabbits.

At the beginning of the fourth month, the first pair reproduces again, but the second pair is not mature enough – so there are three pairs. In the fifth month, the first pair reproduces and the second pair reproduces for the first time, but the third pair is still too young – so there are five pairs. The mating ritual continues ad infinitum. These are mathematically idealised rabbits that never die and always produce a perfect pair of male and female rabbits. What you soon realise is that the number of pairs of rabbits you have in any given month is the sum of the pairs of rabbits that you had in each of the two previous months. So the sequence goes – 1, 1, 2, 3, 5, 8, 13, 21, 34, 55, . . .

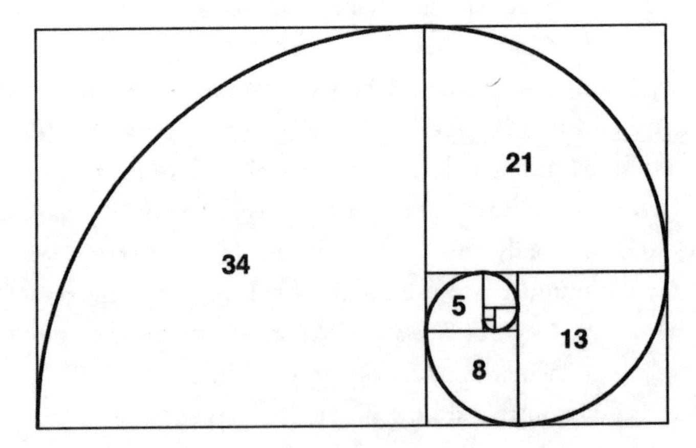

Fibonacci's curious riddle about rabbits introduced mathematicians in Europe to this sequence of numbers for the first time. But once they were on the scene, people began to understand that it wasn't just rabbits that liked these numbers. Wherever there was growth, these numbers seemed to be part of the code underpinning that growth.

The fact that each consecutive number emerges from adding the previous numbers hints at how growth is naturally built into this system of numbers. But it is when one views these numbers geometrically that one understands they might be a natural set of numbers for the organic building of structures.

If we take a 1×1 square, then the system only knows about this structure. So the best it can do is to replicate this structure. But now we have two 1×1 squares. Put these together and the system learns about the number 2. It then produces a 2×2 square that it adds to the structure. But this now generates a 2×3 structure. So the system has generated the number 3. By adding a 3×3 square to the evolving shape we have grown the number 5. By continuing to add squares corresponding to the new numbers that have been made, the structure grows. The new squares create larger and larger rectangles. But, conversely, we see how naturally a rectangle of Fibonacci dimensions can be divided into smaller cells.

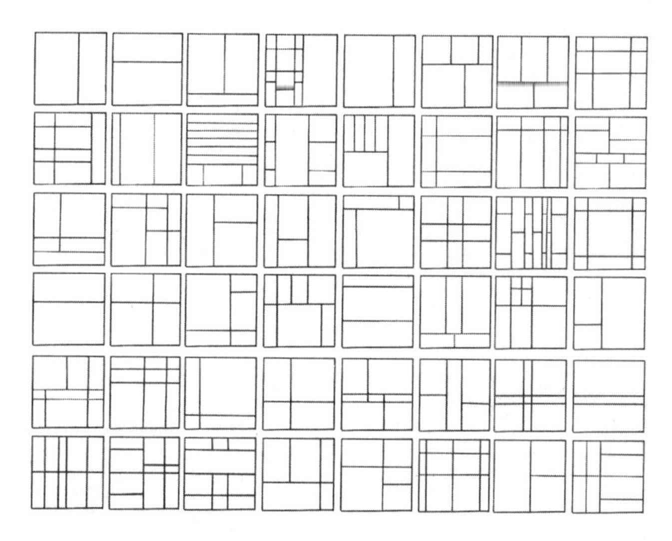

This geometric characteristic is the appeal of this sequence of numbers for an architect or designer. They are a sequence of dimensions that produce building blocks that fuse together very naturally. In Le Corbusier's notebooks, you can see the architect playing around with the huge variety of ways these building blocks can be arranged. There is a plethora of blueprints based on these numbers for possible layouts of rooms in his buildings.

Le Corbusier wasn't the first to tap into the power of mathematical proportions as a blueprint for the buildings we inhabit. Probably the master of proportion in architecture is the sixteenth-century Italian Andrea Palladio.

Frozen music

When you enter a Palladian villa, there is something which just feels perfect about the space. The reason is that Palladio was tapping into dimensions that have a universal mathematical resonance. He would often build his villas around rooms that were perfect cubes. The width, depth and height of the room perfectly matched. He often liked to put two cubes together to create a room whose footprint was in a 1 to 2 proportion. But he also enjoyed slightly more interesting whole number proportions: 2 to 3, 3 to 4, 3 to 5.

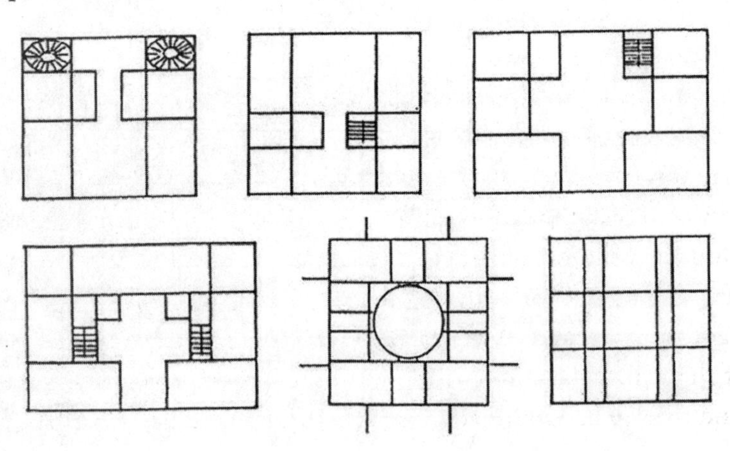

Like Le Corbusier, Palladio littered his notebooks with experimental diagrams exploring the mathematical possibilities of buildings constructed from rooms in whole number ratios. In contrast to Le Corbusier's doodles, Palladio's sketches are highly symmetrical. Symmetry was an important ingredient for the buildings of the sixteenth century, something that Le Corbusier enjoyed disrupting in his architecture for the twentieth century. This asymmetry is encapsulated in Le Corbusier's iconic Modulor Man, who has one hand raised above his head, the other at his side.

These whole number proportions are not just fundamental to creating Palladio's buildings, they are the building blocks of the entire musical world. It is one reason why Palladio's architecture is often referred to as 'frozen music'.

The discovery that these mathematical ratios are key to musical harmony goes all the way back to Pythagoras in the sixth century BCE. Legend has it that Pythagoras was passing a blacksmiths' forge one day and was struck by the beautiful sound that two hammers made as they hit the anvil together. He wondered what it was about the sounds of these hammers that made them so aesthetically pleasing to the human ear. When he tried to recreate the notes on a string instrument, he discovered the answer. It was actually mathematics that we were responding to.

When he played a note on an open string, and then halved the length of the string, he got a note that was higher and yet mysteriously sounded very similar to the lower note. We now call this interval made up of two notes in a 1 to 2 proportion 'the octave', and give these two notes the same name because they sound so similar. It's the feeling of arriving where you started that Bach exploited in the Goldberg Variations when the two voices in the canons drift apart in each subsequent canon yet eventually seem to come together again at the octave.

It's when Pythagoras divided the string at a third of its length that he got really excited. When played against the note he created from half the string, the combination was beautifully harmonic. It was the sound that he had heard coming from the two blacksmiths' hammers.

The key to this harmony was two notes whose wavelengths were in a 2 to 3 relationship. This harmonic pair we call a 'perfect fifth', and it is the beginning of all music across the planet. It doesn't matter what culture or country you come from, everyone recognises this interval as harmonic. We are all responding to the universal nature of the mathematical blueprint that underpins it.

Pythagoras didn't stop there. If you start with one note, you can add a new note in this 2 to 3 relationship to create a harmonious pair. But now take this second note and add a new note above that, creating a new harmonious pair. You can keep on doing this, creating ever more harmonious pairs. However, something rather magical happens when you've done this 12 times. The new note that appears sounds remarkably like the note you started with, just seven octaves higher. And this return to the first note after 12 perfect fifths is the reason that the musical octave is divided into 12 notes, known as the chromatic scale. If you look at the notes on a piano, there is a pattern of 12 black and white keys that is repeated across the keyboard. These and the corresponding octaves are the notes that Pythagoras' cycle of fifths naturally picks out.

This explains why it isn't arbitrary that music is divided into 12 notes but a mathematical inevitability. We couldn't, for example, have gone decimal and divided the octave into 10 notes. It just wouldn't work harmonically. The reason is that, every time we add a new harmonic note, we are multiplying the frequency of the previous note by 3/2. But if we multiply 3/2 together 12 times it is remarkably close to multiplying 2 together 7 times:

$$(3/2)^{12} \sim 2^7.$$

Doubling the frequency of a note produces the same note an octave higher. If you keep adding more perfect fifths, you don't really get any new notes, just copies of the ones you've already generated. It's this mathematical coincidence that 3^{12} is close to $2^{7+12}=2^{19}$ that determines the 12 notes of the musical universe. But notice that 3^{12} is not the same as 2^{19}. One is odd and one is even. Music is not mathematically perfect. Our ears can't really hear the difference between these two numbers, but the fact that there is a difference causes some interesting issues when it comes to tuning a piano.

Even though other musical instruments, such as those used in the Indonesian gamelan, divide the octave further, the backbone of this 12-note division is universal across all cultures. Even more evidence for the universality of this mathematical foundation of music is the fact that even birdsong features the same harmonic division.

When I listen to the blackbird and the song thrush that sing to me every morning and evening from the tree outside my bedroom window, the sound very deservedly earns the title of 'song'. And it is certainly a complex song. Partly that is because of the speed at which the notes fly about. It's almost as if my brain can't keep up with the patterns, while the mate or rival that the bird is singing to has the equipment to decipher this complex sound. That the speed of delivery was part of the problem was revealed to me when the viola player from the string quartet I play with recorded the birds one evening during a break in rehearsals at my house. He then used some technology to slow the sound down without altering the pitches. Taken at an 8th or 16th of the actual speed, the song sounded much more human. But were the pitches matching the mathematics of the music we were playing?

I discovered that the question had been addressed in a fascinating paper published in 2014 in the *Proceedings of the National Academy of Sciences* in America. I almost wondered if the paper was an April Fool when I read the name of the lead author: Emily Doolittle. Dr Doolittle, talking to the animals! Well, if not talking, she was listening to the animals. A hermit thrush in particular. Doolittle is a composer, but she teamed up with scientists to explore the pitches that the thrush uses in its song.

After analysing 114 song types from 14 male hermit thrushes, they were excited to discover that the notes the thrushes were choosing were related by the same whole number ratios that Pythagoras revealed are the key to human music. It explains in part why Messiaen, who was obsessed by birdsong, was able to transcribe the song of the nightingale and the blackbird into the music that opens the *Quartet for the End of Time*. Without this common mathematical blueprint, it would have been impossible.

From simplicity to irrationality

Palladio's rooms, in their 1 to 2, 2 to 3, 3 to 4 relationships, had captured in the dimensions of his buildings the combinations of notes that sound pleasing to the human ear. As Palladio wrote in 1567: 'The proportions of the voices are harmonies for the ears; those of the measurements are harmonies for the eyes. Such harmonies usually please very much, without anyone knowing why, excepting the student of the causality of things.'

Pythagoras was so excited by the discovery that harmony was just maths in disguise that it led to his belief that the entire universe was based on blueprints that consisted of whole number ratios. But there was another mathematical blueprint Palladio enjoyed using which revealed that Pythagoras' blueprint for the universe was too simplistic.

This second blueprint that Palladio loved to exploit in his buildings was a rectangle possessing the magical property that if you divide it in half, then the smaller rectangle that emerges has exactly the same proportions as the original shape. This rectangle worked beautifully as a blueprint for rooms inside a building because you could divide the room in two and the smaller rooms have the same proportions as the larger room, just at half the scale.

This scaling might remind you of another place where proportions are preserved when dividing something in half. Take an A4 piece of paper and fold it in half and you get an A5 page. Same proportions, just half the size. The A-series of paper sizes is designed with this property in mind. Starting with an A0 sheet, every time you divide the paper, the page shrinks by half, but the proportions are the same as the original. The idea for paper sizes with these properties was first proposed in 1786 by the German scientist Georg Christoph Lichtenberg and was briefly adopted as a new paper standard during the French Revolution. But it didn't really catch on globally until the middle of the twentieth century.

How do you create a rectangle with this magical property? The key, of course, is mathematics, and the role of an important sort of number which rocked the mathematical community that discovered it during Pythagoras' day. If the short side of a page is 1 unit in length, and the long side is A units in length, then what value of A will ensure that when I fold the paper in half the proportions remain the same? The smaller piece of paper now has a short side equal to A/2 and a long side 1 unit in length. If the proportions are the same, then A:1 has to be the same as 1:A/2. This becomes the equation:

$$\frac{A}{1} = \frac{1}{A/2}$$

Rearranging this reveals that A is a number whose square is 2. Pythagoras had already encountered the challenge of finding a number whose square is 2 when he was investigating the simple geometric properties of one of the fundamental shapes that formed part of his blueprint for the universe: the square. To calculate the distance in a square of 1 unit sides from one corner of the square to the opposite corner, Pythagoras applied his famous theorem about triangles. It told him that the square of the distance is equal to the sum of the squares of the sides of unit length, namely $1^2+1^2=2$. So the distance across the square had a length whose square had to be 2. Given his belief that the universe was made from whole number ratios, the hunt was on for a fraction that had a square equal to 2, a fraction that could measure this length across the diagonal of the square.

But when one of Pythagoras' followers, Hippasus, came up with a proof that no fraction could have a square equal to 2, it completely destroyed this blueprint of a universe built from whole numbers. The myth goes that Pythagoras ordered his followers to drown Hippasus in an attempt to suppress this discovery. But that's the power of mathematical proof. Once the truth is established, it will survive any human attempts to fashion the universe otherwise. Hippasus' proof led to the revelation of a whole series of new numbers, which we now call 'irrational numbers' because they are not ratios. If you try to express them as decimal numbers, then their decimal expansion goes on to infinity and never repeats itself. The golden ratio, which we will encounter in the next blueprint, is another example of an irrational number.

Hippasus' proof of the irrationality of the square root of 2 was the second mathematical story that G. H. Hardy presented in his book *A Mathematician's Apology*, which had captured my teenage imagination. It was not so much the result that excited me but the way it was arrived at. The final chord of a piece of music is really only exciting because of the journey that you've been on to reach that climax. The proof of the irrationality of the square root of 2 captures much of the aesthetic and narrative quality that for me is important in mathematics. So just as Hardy tried to share the story with readers, I want to try to do the same and communicate something of the drama that it entails. You can skip forward if you are feeling too daunted by the challenge, but my hope is that you might feel brave enough to have a go. Open your mathematical eyes and let's dive in.

I want to prove to you that it's impossible for a fraction to have a square that is 2. The proof employs the same powerful narrative device we saw at work in the first blueprint, in the proof that there are infinitely many primes: suppose the opposite. Let's suppose there is a fraction that has a square equal to 2 and explore the consequences of such a belief. If we do that, then we discover that there is a number that is both odd and even at the same time. That is clearly absurd. That would contradict mathematics. A number is either even, divisible by 2, or it is odd, not divisible by 2. It can't be both. If I can cook up this contradiction, it will show that our belief that there is a fraction whose square is 2 is wrong. As Sherlock Holmes famously declared: 'When you have eliminated all which is impossible, then whatever remains, however improbable, must be the truth.'

Our task, therefore, is to show why a fraction whose square is 2 would imply the existence of a number that is simultaneously odd and even. Let p/q be the fraction that we believe exists whose square is 2. We can make some assumptions about this number: for example, one of the numbers p or q we can choose to be odd, because if they were both even then I could cancel the 2s that divide both top and bottom of the fraction, reducing it to a form where one number at least is odd. So one of the numbers p and q is odd. That is the first part of our task. We will then prove that this odd number is actually even. That will follow from the fact that this fraction has a square equal to 2:

$$(p/q)^2=2.$$

We are going to do a bit of algebra, but nothing too frightening, I hope. First of all, we can rewrite the equation above as:

$$p^2/q^2=2.$$

This is just a consequence of how we multiply fractions: $a/b\times c/d=(a\times c)/(b\times d)$. Now we are going to multiply both sides of this equation by q^2. On the left side, this means we can cancel the q^2. So we get a new equation:

$$p^2=2\times q^2.$$

The number on the right is even because it is divisible by 2. An odd number multiplied by an odd number is odd. So an odd number squared must be odd, which means p can't be odd, so it must be even. But remember that we know that one of p and q is odd. So it must be q that is the odd number. But let's prove why q is also even. Since p is even, then $p=2\times n$, where n is some new whole number. But $p^2=(2\times n)\times(2\times n)=4\times n^2$. Let's put this back in the equation we have above:

$$4\times n^2=2\times q^2.$$

We can cancel a multiple of 2 on both sides to get:

$$2\times n^2=q^2.$$

Now we have our moment of triumph. The number on the left is even. But that means that q must be even because only even numbers have squares that are also even. So q is both odd and even at the same time. This is the consequence of starting with the belief that there is a fraction p/q whose square is 2. That belief must be wrong. There is no such fraction.

This is called a 'proof by contradiction'. There is something about the dynamic of the logical argument that almost feels like a battle between the two sides of the equation, as if they are trying to outdo each other by coming up with ever more multiples of 2. This dynamic quality of the proof was actually the inspiration for a piece of contemporary dance that I devised with New Zealand choreographer Carol Brown. I even ended up performing the proof with a dancer at the Trinity Laban Conservatoire of Music and Dance in London. It was an appropriate venue for a mathematically inspired dance because, as I will explain in a later blueprint, the choreographer Rudolf von Laban – after whom the Conservatoire is named – based his whole theory of dance on the importance of mathematical geometry to a dancer's movements.

The score for the piece of choreography we devised followed the lines of the proof in an almost combative argument between the two dancers. It reminded some members of the audience of Indian Kathak dancing, a style which possesses some things in common with Laban's ideas. The person I danced with was struck by how easily I remembered the moves, but for me I was just following the proof in my head. The logic of the proof implied an inevitability to the moves we were making.

The discovery that the square root of 2, this length across the diagonal of a unit square, cannot be represented as a ratio of whole numbers has deep implications for the idea of the universe as a physicalised piece of mathematics. If we really are living in Plato's cave, watching a shadow on the wall of this pure abstract world of mathematics, then this theorem implies that the shadow is a poor, bastardised version of the mathematics.

Modern physics now believes that space is quantised, that it is made up of indivisible units of space. So if I draw a square, then the side of the square consists of M units of indivisible space. Similarly, the diagonal across the square consists of N units of indivisible space. Then $M^2+M^2=N^2$ by Pythagoras' theorem, which in turn implies that $2=(N/M)^2$. That would mean the square root of 2 could be expressed as a fraction, but the Pythagoreans proved that is a mathematical impossibility. The way to resolve this is to recognise that you can never physically make a perfect diagonal across a square. Space must bend or fold in order to fit the length across a square. The idealised mathematics of a perfect square with a diagonal drawn across is actually a physical impossibility. Quantum physics denies the possibility of forming this mathematical blueprint in reality. It remains in Plato's abstract realm, whose shadow on the cave wall is but a poor copy.

It's interesting that this inability to write the square root of 2 as a fraction has actually had significant implications for music too. One of the reasons that the issue of tuning a keyboard instrument like the piano or harpsicord has been so contentious ever since Bach's day is because if you try to use the harmonious notes that Pythagoras discovered to do so, then you run into difficulties.

Pianos nowadays are commonly tuned according to a system known as 'equal temperament', which means that the ratio between the frequency of two consecutive notes on the piano is always the same. So the frequency of this higher note is a multiple X of the lower frequency. As we climb through the 12 notes of the chromatic scale, the frequency gets multiplied each time by X. Once we have done this 12 times, we reach the octave. But we know that this frequency is double the frequency we started with, which means that $X^{12}=2$. X has to be the 12th root of 2. But since the square root of 2 is not a fraction, this implies in turn that there is no fraction whose 12th power is 2. This means that X is an irrational number. But Pythagoras makes music that is all in ratios of 3 to 2; it's all fractions. This incompatibility between equal temperament with its irrational numbers and Pythagorean scales made up of fractions was at the heart of a long debate on how you should tune instruments which raged in the sixteenth and seventeenth centuries.

It was during this period that square roots gave rise to the creation of yet more interesting numbers. The ancient Babylonians had come up with a clever formula for solving quadratic equations involving x^2. Attention shifted in the sixteenth century to the challenge of finding a way to solve more complicated equations involving x^3, known as cubic equations. Early attempts to come up with a formula threw up the challenge of taking the square root of a negative number. Just like the square root of 2, there didn't seem to be any numbers that solved this problem. At least the square root of 2 existed as the distance across the square. The square root of a negative number didn't have a comparable interpretation. Perhaps no such numbers existed. That was the stance taken by mathematicians when the challenge first appeared on the scene.

But one mathematician, Gerolamo Cardano, started to experiment with a mathematics where there were numbers whose square was negative. His investigations began to open up a whole new chapter in mathematics. Descartes had thought the whole idea absurd and coined the term 'imaginary numbers' as a derogatory name for these ridiculous numbers. But the name was rather apposite because these numbers could not be seen or measured, but only conjured up by the imagination. Imaginary numbers are now at the heart of modern mathematics, and of physics too since they are the perfect language for handling the quantum world. The power of the mathematical imagination is that it can lead us to new ways of navigating the world in which we live, just as the best art can do. But for me the pleasure is not so much in the utility that these ideas bring but in being taken on extraordinary intellectual and emotional journeys of the mind.

Like Pythagoras, Le Corbusier was somewhat in awe of the power of mathematics as a blueprint for everything from buildings to the universe.

'Nature is ruled by mathematics, and the masterpieces of art are in consonance with nature; they express the laws of nature and themselves proceed from those laws.' Le Corbusier wrote extensively about the connections between nature, mathematics and the creative arts. Both the mathematician and the artist are responding to the natural world around them so it was inevitable le Corbusier believed that the formulas of mathematics had a role to play in the work of the artist.

As well as mathematics, music was an important part of Le Corbusier's world. His mother, who remained an important force throughout his life, was a pianist. Le Corbusier believed that just as Pythagoras had used mathematics to divide up the continuous world of sound into discrete frequencies from which music emerged, the infinity of space should have a natural scale or system of division that brought harmony to the measuring of the universe. This was his goal. To be the Pythagoras of continuous space. He believed that, in order to maintain harmony in a world becoming increasingly mechanised and in which mass production was beginning to dominate, a standard measure attuned to nature was more important than ever. Man-made items retained their connection to the body, but machines would need guidance in modulating their output with the sensitivities of humanity. His red and blue series were his proposal for a geometric scale that divided space just as Pythagoras' musical scale had divided sound.

Indian origins

Although Le Corbusier's red and blue series seemed innovatory when they were introduced, they were really just numbers that had been known in Europe since the thirteenth century. But it turns out that Fibonacci too had recreated a series of numbers that had been known for centuries before his revelation that they explained population dynamics in rabbits. Very often the mathematics comes first and then its application as a blueprint for the arts is second. But when it comes to the Fibonacci sequence, the numbers were actually discovered by musicians and poets exploring rhythm before they were connected to mathematics and nature.

During the pandemic lockdown in 2020, I decided to distract myself by learning a new language. I was writing a new play about one of my mathematical heroes, André Weil, and I discovered that he had dedicated his spare time to mastering Sanskrit. So I thought that following in his footsteps would be a great way to get in character. I would start every morning by writing out a passage from the Bhagavad Gita, the 700-verse Hindu scripture that forms a part of the epic *Mahabharata*, and then I would try to decode its meaning.

I discovered through my morning musings that Sanskrit words are made up of two sorts of sounds. A syllable in a word can be *laghu*, meaning 'light', or it can be *guru*, meaning 'heavy'. The syllable count is one way to analyse a line of poetry, but there is another approach, which records the amount of time each line takes. When analysing language, the idea of a basic unit of time is often called a 'mora'. The term has its origins in the Latin word for 'a delay' or 'a duration'. What is important in Sanskrit is that a heavy syllable is worth two morae while a light syllable is worth one mora.

I am going to use a dot (·) for a short or light syllable which is one mora long. I'll use a dash (–) for a long or heavy syllable which is two morae long. So each possible template is going to look like a piece of Morse code.

In poetry like that of Shakespeare, we are used to the idea of quite a rigid periodic pattern that is used repeatedly, like the iambic pentameter we encountered in the blueprint on primes. Classical Sanskrit verse on the other hand is characterised by metrical templates that are much more aperiodic and varied throughout the poem. There is a vast range of different templates that are used, totalling 600 or 700 recorded patterns of rhythm.

Sanskrit poetry divides into forms which keep the number of syllables (*varnas*) constant in each line, known as *varnavritta*, or those which keep the total morae length constant, known as *matravritta*. The *matravritta* forms are less common but are often found in texts of Indian philosophy. For example, the *arya* verse form has four phrases that consist of 12, 18, 12 and 15 morae respectively. It was employed by the great poet Kalidasa in his fourth-century CE Sanskrit play *Shakuntala*:

> आ परितोषाद्विदुषां न साधु मन्ये प्रयोगविज्ञानम् ।
> बलवदपि शिक्षितानामात्मन्यप्रत्ययं चेतः ॥ २ ॥
> *āparitoṣād viduṣāṃ*
> *na sādhu manye prayogavijñānaṃ*
> *balavadapi śikṣitānām*
> *ātmany apratyayaṃ cetaḥ*
> – · · | – – | · · –
> · – · | – – | · – · | – – | –
> · · · · | · – · | – –
> – – | – – | · | – – | –

I do not think a play is performed well until it satisfies a learned audience; even someone highly trained in performance lacks self-confidence.

Notice how random the combinations of long and short beats are in this Sanskrit verse compared to the kinds of metre that you find in the poetry of Shakespeare, for example. This is a culture that is exploring the limits of metrical possibilities.

It is not just literary ideas that the Indians were capturing in verse. Mathematics in India was also written in poetry and the sixth-century CE mathematician Aryabhata would use a *matravritta* form to record his discoveries. These included a very accurate estimate for π as well as formulas for adding up series of square and cubic numbers.

It is the *matravritta* forms that are interesting in respect of the Fibonacci numbers because poets became intrigued by the question of how many different templates might exist for a line containing, say, 16 morae.

Let me start by exploring all the possible templates for lines with four morae:

– –
– · ·
· – ·
· · –
· · · ·

Five possible templates. What if I now wanted to calculate the lines with five morae? I could take all these templates and simply add another short beat to the end of them. Or I could take the lines of three morae and add a long beat to the end of them. There are three of these:

– ·
· –
· · ·

So the templates for five morae are obtained by adding a long beat to the end of the three templates with three morae and a short beat to the end of the five templates with four morae, making a total of eight templates. The Fibonacci rule has emerged as the secret to generating ever more complex templates for the Sanskrit poets. To discover the number of possible templates with 16 morae, we have to find the 16th number in the sequence that starts:

1, 2, 3, 5, 8, 13, . . .

Notice that unlike Fibonacci's rabbits, which began with a repeated 1, the number of templates for rhythms only has a single 1. By continuing the rule of adding the two previous numbers to generate the next number in the sequence, the algorithm reveals that there are 1,597 possible templates with 16 morae.

When did the Indian poets move from just experimenting with different templates to a rigorous mathematical analysis of what might be possible? When did metrical art turn to metrical science? Acharya Pingala can probably be credited with some of the earliest known mathematical explorations of metrical possibilities in poetry. His book, the *Chandas-Shastra* – sometimes known as the *Pingala-Sutras* – dates back to at least the second century BCE and contains a series of rather cryptic sutras about how to generate different templates. Subsequent commentaries, in the tenth century CE, decoded Pingala's analysis and credited him with understanding the Fibonacci rule as the secret algorithm for calculating the number of templates as the lines got longer.

Pingala also covered verse that counts syllables rather than morae. Again, a syllable might be heavily or lightly stressed, as in iambic pentameter, which consists of five sets of sounds corresponding to dee-*dum*, or a light stress followed by a heavy stress. The possibilities now correspond to powers of 2, because in a line with eight syllables, for example, there is a choice of a heavy or a light stress in each position. This makes $2^8=256$ different possibilities. Some have credited Pingala with the first use of the symbol for zero because he indicated a light beat with a 0 and a heavy beat with a 1. Each template then resembles a binary number. Iambic pentameter is 0101010101. It is probably going too far to say Pingala also discovered binary numbers since his 1s and 0s are just simple representations of rhythms rather than actual numbers.

Pingala was clearly a mathematician at heart because he also invented a fun algorithm or mnemonic to help the reader remember all the $2^3=8$ different possible three-syllable feet made up of heavy and light syllables. It's the nonsense word:

yamātārājabhānasalagāḥ.

In Pingala's binary representation of stressed syllables, the word appears as:

0111010001.

If you take all the three-syllable combinations that make up this ten-syllable word, they correspond to the eight different combinations that you can make:

011, 111, 110, 101, 010, 100, 000, 001.

There are other Indian thinkers whose names arise as the possible discoverer of the Fibonacci sequence. They include Acharya Virahanka, who lived between the sixth and eighth centuries CE, and who wrote the following rule for obtaining the new templates from old: 'The variations of two earlier metres being mixed, the number is obtained. That is a direction for knowing the number of variations of the next *matravritta*.'

The twelfth-century Jain philosopher Hemachandra also wrote a study of poetic forms, the *Chandonushasana*, which documented many different metrical combinations. Here is his example of a 16-morae rhythm called *Suddhavirat*:

vis vam tis tha ti kuk si ko ta re
– – – . . – . – . –

The universe rests in the cave of the womb.

His treatise also included the mathematical description of the algorithmic rule for generating rhythms and there has been quite a drive in recent years to have his name replace Fibonacci in the history books.

The role of Fibonacci numbers in poetry has had a resurgence in modern times with the proposal of a new poetic structure called 'the Fib'. Like a haiku, the form is defined by the number of syllables in each line. The haiku is a Japanese form with five syllables in the first line, seven in the second and then five again in the third. With a total of 17 syllables, it is dripping with indivisible prime numbers, and yet the symmetry of the structure creates a satisfying counterbalance to the disruptive quality of the indivisibility.

With a Fib, the syllables of each line in the poem correspond to the Fibonacci numbers. It starts off with a rather staccato sound. A monosyllabic word in lines one and two, then a two-syllable word or phrase in line three. But then, gradually, the exponential growth of the

Fibonacci numbers allows for more fluid lines to emerge: three syllables, five syllables and finally eight syllables. The poem has a total of 20 syllables in six lines.

The form had been around for many years, but it received a burst of attention after a blogpost in 2006 by the American writer Greg Pincus, in which he illustrated the form with the following Fib:

One
Small,
Precise,
Poetic,
Spiraling mixture:
Math plus poetry yields the Fib.

The post went viral: the comments section below it began to fill with everyone's experiments in the form, and the story was picked up by major newspapers.

There is something very inviting about the challenge of writing a Fib that encourages first-time poets to start composing. This is often the power of a good blueprint: the structure provides a helpful scaffold to explore one's creativity without being frightened off by the blank page. As David Usborne in the *Independent* newspaper commented: 'There is the danger, of course, that once you start, devising Fibs will become as addictive as crossword-filling or Sudoku-solving.'

The attraction of the form itself probably owes something to that sense of natural growth that the Fibonacci numbers encode. Each line grows organically from the sound of the two previous lines. Of course, you could continue the form beyond the sixth Fibonacci number, but then there is the danger of the poem descending into prose rather than poetry.

When I was asked some years ago to speak at the Ledbury Poetry Festival in England about the connections between mathematics and poetry, I thought that the Fib was a fun form to include in my discussions. In preparation for the talk, I offered a challenge to my followers on Twitter: tweet me a Fib. I thought the restriction of 140 characters that existed at the time would make a tweet a great place to record a Fib. My favourite Fib of the lot came in from @benbush:

Tweet
Tweet
Marcus
Here's my fib
(An unwise ad lib?)
Wait: fib? On Twitter? I'm confused
How many of my 140 have I used.

I thought it was particularly nice because Ben pushed the form naturally to a seventh Fibonacci number, with a last line of 13 syllables.

Musical rhythm

It wasn't just the poets in India that were excited by generating interesting rhythmic templates. Indian music is infused with complex rhythmic patterns performed on the tabla, a pair of hand drums. I was introduced to the exciting mathematics of the tabla when I advised the theatre company Complicité on their show about mathematics, called *A Disappearing Number*. At the heart of the play was the collaboration between G. H. Hardy, the author of *A Mathematician's Apology*, and the Indian mathematician Srinivasa Ramanujan.

The company decided that music was an interesting medium in which to represent the two different cultural approaches to creating mathematics. To explore the way that Indian music is composed, they brought British composer Nitin Sawhney into the project. Born to Indian parents, Sawhney had been brought up learning the sitar and tabla. I was privileged to take part in the workshops he ran during the development of *A Disappearing Number*, and they made me realise how important mathematics and number are to the unique sound of Indian music.

The basic element of tabla playing is the *tala*, or beat. *Tala* is a Sanskrit word meaning 'being established', and this beat is fundamental to creating the timescale of the piece. It is combined with the *raga*, or melodic component, which constitutes the heart of Indian music. Although Indian music involves a lot of improvisation, there are still rigid structures that underly the piece that the players agree on before they start playing.

Because music is more tightly restricted by time than poetry, the precise length of the long and short beats is even more important. If a piece has eight beats left, then a tabla player will have a choice of different rhythms to fill the remaining time – but the long and short beats must add up to eight. So the number of possibilities available to the tabla player is the eighth number in the sequence that the poets had discovered, which means there are 34 different rhythms that the tabla player can use.

The word *tala* can also be used to describe these rhythmic templates which a tabla player might employ. They can vary from *talas* that are as short as three beats to complex ones spanning 128 beats. Very often these templates are repeated several times in succession. They form the backbone of the rhythmic structure of a piece. Common *tala* lengths in Indian music are 7, 8, 10 and 16 beats. The 8 and 16 are rhythms that will often be familiar to Western ears, but the use of a prime number like 7, or the two lots of 5 in the 10 beat *talas*, are often quite unsettling, as if a beat is missing or an extra beat has been added. It is exactly the same trick that Björk or Radiohead exploit in their complex rhythmical songs.

A whole variety of sounds can be made on the tabla, and there is a mnemonic system known as *bol* which is used to name these different effects. A particular *tala* on the tabla can then be translated into a spoken phrase, and this is often how the *tala* is learnt. You can sometimes even hear the *talas* being spoken by the non-tabla players during performances. For example, here is the mnemonic for a common seven-beat *tala*:

Tin Tin Na Dhin Na Dhin Na.

This 7 is actually thought of as being made up of 3+2+2. The 3 and 2 beat sequences are used as atomic *bols* to build the other more complex rhythms. This additive quality of Indian rhythms is very typical.

There are some fun mathematical games that tabla players enjoy. For example, the *tihai* is a rhythmic sequence that is repeated three times at the end of a piece or phrase. It often includes a pause or rest between each repetition. But the idea is that the end of the *tihai* should land on a significant beat in the piece. So, the piece might be broken up into groups of 16 beats. This is very typical in many musical cultures. It's four lots of four beats. But what makes Indian music so distinctive is the way that the *tihai* is used to disrupt the simple structure. A *tihai* might consist of a five-beat rhythm repeated three times with a single beat separating them:

$$5+1+5+1+5=16+1.$$

The last beat of the *tihai* would therefore land on the first beat of the new round of 16. But the mix of fours against this repeated five causes an exciting tension that is very typical of Indian music. It starts to become hard to hear where the main beat is when a tabla player knocks out these *tihais*.

In Western music, tension is often created by harmonic chord progressions, but this is not one of the core ingredients in Indian music where the *raga* corresponds to the melody and the *tala* is the rhythm. The harmony is missing. So these *tihais* are used in place of a harmonic chord progression to create that sense of tension and resolution.

Given that you can vary the length of the pause and the length of the repeated rhythm, plus the fact that the overarching structure may employ different lengths, you can see how mathematics will necessarily feed into the rhythmic choices of the tabla player. Essentially there are two different journeys going on simultaneously which feel disconnected until they suddenly come together on the satisfying common beat. It resembles Messiaen's idea of using prime numbers to keep the harmony and rhythm out of synch. Messiaen's explorations into Indian rhythms were probably the inspiration for his use of primes in the *Quartet for the End of Time*.

In poetry, rhythms are made up of long and short beats, or of heavy and light stresses. But music has the potential to be more varied. And so the question arose of mixing up rhythms composed of long, short and a third length in between those two. The thirteenth-century musicologist Sharngadeva was one of the most influential writers in the medieval period on Indian classical music. His treatise, the *Sangita-Ratnakara*, known in English as *Ocean of Music and Dance*, is revered by musicians across India. In the fifth chapter, the author makes a list of 120 fundamental *talas* that can be made up from these three different lengths of note. They are known as the *deci-talas*.

Deci, or *desi*, is derived from the Sanskrit word meaning 'from the country', and these *deci-talas* were gathered from the South Asian region made up today of India, Bangladesh and Pakistan. *Deci* is still used today in modern parlance to indicate people from this region. For example, MTV set up a dedicated MTV Desi station targeted at the *deci* community in America to meet the growing demand for popular programming from South Asia. Some regard it as a positive term, but others have criticised it for homogenising a whole region when it actually possesses a broad range of cultural traditions.

The 120 *deci-talas* are not a comprehensive list of all possible rhythms, but rather a compilation of greatest hits. It raises the fascinating question, though, of how many possible rhythms you can make with notes that are one, two or three beats long. The rule for generating rhythms of length N with one or two beats is that you add the two previous numbers together. This corresponds to adding a short beat to the N−1 length rhythms and a long beat to the N−2 length rhythms.

If we now have rhythms with the possibility of a three-beat rhythm then the rule is simply that we add together the previous three numbers in the sequence. Let me denote by R(N) the number of rhythms of length N. The sequence starts with:

$R(1)=1$ There is one rhythm possible of length 1

$R(2)=2$ There are two rhythms possible of length 2: 1+1 and 2

$R(3)=4$ There are four rhythms possible of length 3: 1+1+1, 2+1, 1+2 and 3.

But now we can use our rule for generating longer rhythms from this point.

$R(4)=R(3)+R(2)+R(1)=7$
$R(5)=R(4)+R(3)+R(2)=13$
$R(6)=R(5)+R(4)+R(3)=24$
$R(7)=R(6)+R(5)+R(4)=44$
$R(8)=R(7)+R(6)+R(5)=81$
$R(9)=R(8)+R(7)+R(6)=149$
$R(10)=R(9)+R(8)+R(7)=274$.

These are sometimes called the 'Tribonacci numbers', because the rule that is used to generate them has length 3. Intriguingly, this sequence has a role to play in the natural world too. It is mentioned in Charles Darwin's *On the Origin of Species* in relation to the number of pairs of elephants that are born in each new generation, reminiscent of the Fibonacci numbers counting rabbit populations. In Darwin's model, a pair of elephants gives birth to another pair in years 30, 60 and 90, but then stops reproducing and dies. So to calculate the number of new pairs of elephants born in each subsequent 30-year period, Darwin recognised that you just need to add the three previous numbers together. Note that Fibonacci was keeping track of total pairs of rabbits that lived for ever and kept reproducing. Darwin was counting just new pairs of elephants and they stopped reproducing after three generations.

He used this example to illustrate the contrast between the exponential growth of species and the linear growth of resources. He wanted to make an important point about the challenge implied for sustainability given these two different models of growth: 'There is no exception to the rule that every organic being naturally increases at so high a rate, that if not destroyed, the earth would soon be covered by the progeny of a single pair.'

Although he was good at biology, Darwin seems to have got into a bit of a muddle with his maths. Over the six different editions of *On the Origin of Species* published during his lifetime, his calculations using his Tribonacci formula go through a number of variations. In the first edition of *On the Origin of Species*, Darwin stated that after 500 years there would be 15 million living elephants descended from the first pair. This statement was reproduced in subsequent editions until, after the fifth edition, Darwin started receiving letters from various readers unable to reproduce the calculations alluded to in his book. One such letter appeared in *The Athenaeum* in 1869, signed by a correspondent going by the name 'Ponderer', who enquired: 'Perhaps some of your readers will be able to enlighten my dull intellect as to the process of reasoning by which this result is obtained.' His calculations using the Tribonacci rule had led to a very different answer from Darwin's:

> The number of elephants alive at that time would be 42,762 pairs, that is, 85,524 elephants, less the number that would have died by reason of their age. But Mr. Darwin says that there would be fifteen millions. On what does he base his calculation?

Darwin realised something was amiss. He got his fifth child, George, who was a mathematician, to redo the calculations. These yielded the result that after 25 generations, each 30 years long, there would be 5,111,514 elephants. Darwin was clearly horrified by the error. In a letter from April 1872, he blamed it on the fact that 'I got some mathematician to make the calculation, and he blundered and caused me much shame'.

But looking back at the original manuscript of *On the Origin of*

Species, we find exactly this number. It seems that Darwin was experimenting with periods of 20 years between births rather than 30, which would have given the 500 years he talked about. He also experimented with elephants giving birth to four generations, rather than three, which results in the 15 million figure. His rush to get his book published resulted in him cutting corners and making a total hash of his example. Eventually, by the sixth edition that appeared 13 years later, he had corrected the passage.

Once again we see the triangle of mathematics, nature and art at work. The Tribonacci numbers are the blueprint for both Darwin's elephants and the complex rhythms considered by the Indian musicologist Sharngadeva. The list of 120 *deci-talas* documented in the *Sangita-Ratnakara* were the rhythms that Messiaen had come across when he first read about Indian music in the 1920s.

He immediately saw the potential that this list offered for taking his compositional practice in a new direction. 'The list was a revelation,' he wrote. He started copying out the rhythms, exploring them from every angle, returning to them again and again over the years to tease out their hidden meaning.

The Sanskrit names were something of a mystery. But luckily, an Indian friend translated them for him, allowing him to understand not only the rhythmic rules, but also the cosmic and religious symbolism that is contained in each *deci-tala*.

For Messiaen, there was no music without rhythm. Indeed he thought rhythm came before notes as the essential element of music. Rhythm at its heart was mathematical built from the change of number and duration. But he felt that the Western music his generation had been brought up on had got stuck. Rhythm had become nothing more than the simple repetition of a military march.

Messiaen believed it offered the potential for so much more: 'rhythm is an unequal element, following fluctuations, like the waves of the sea, like the noise of the wind, like the shape of tree branches.'

Discovering the *deci-talas* was like stumbling on a rhythmic cabinet of wonders. On the lookout for rhythms with interesting musical and often mathematical properties, Messiaen would plunder them throughout his life. He used three of the *deci-talas* to construct the rhythm sequence that the piano plays in the first movement of the *Quartet for the End of Time*: the *ragavardhana* (or at least Messiaen's augmentation of this rhythm), the *lakskmica* and the *candrakala*.

These rhythms are interesting in part because the first two have prime numbers hidden inside them. The *ragavardhana tala* (number 93 of the 120) is a rhythm that lasts for 19 units of time, while the *lakskmica tala* (number 88 of the 120) lasts 17 units of time. Many of the *deci-talas* that Messiaen enjoyed have an interesting telescopic quality to them, where the rhythm gets repeated but slowed down or speeded up. As we shall see later, this is an almost fractal structure, where one sees the same thing repeated at different scales. But Messiaen also loved rhythms that had symmetry. He called these rhythms 'non-retrogradable' and talked of the almost religious quality that symmetry represented for him (I will say more about Messiaen's relation to rhythmic symmetry in the seventh blueprint).

The other quality that Messiaen loved about the *deci-talas* was their elasticity. New rhythms could be generated by elongating or contracting notes. They could be added together to create ever more complex sequences. This is the secret power that the Fibonacci blueprint gives to music. Another composer who particularly enjoyed the elasticity made possible by Indian rhythms is Philip Glass.

Additive music

One two three four
One two three four five six
One two three four five six seven eight.

So begins Philip Glass's game-changing opera *Einstein on the Beach*. An onslaught of mathematics. But it's not just in the words. The music too is infused with an exciting cocktail of mathematical ideas which, like Messiaen's, have their origins in India.

I was on a fascinating panel with Glass at the New York Science Festival some years ago, exploring the different styles of thinking required for doing science and art. Glass didn't really believe there was a difference. He has often talked very openly about the way he's used mathematics as a powerful tool for generating musical effects. When Robert Wilson, Glass's collaborator on *Einstein on the Beach*, asked him how he composed, he gave Wilson a mathematical equation. The opera Glass wrote clearly has a very mathematical theme at its heart, but it is embedded in the music too, which was constructed along numerical lines employing a technique that Glass called his 'additive process'.

This process has a lot in common with the recursive nature of the Fibonacci series. It is what makes Glass's music so recognisable and so revolutionary at the same time. The idea is to take the seeds of a melody and then, as the piece evolves, grow them by adding beats or notes or fragments of other phrases to the original theme. It is an almost algorithmic mechanical process, which might be alienating, but the art is to choose seeds and algorithmic rules that combine to draw the listener into the evolving sound world.

One of the clearest demonstrations of this technique at work can be found in a very early piece by Glass called *1+1*. Even its name suggests mathematics at its heart. The piece is one of pure rhythm, something which, as Messiaen articulated, is the foundation of all music.

1+1 is for a single player, who taps out a rhythmic sequence on a tabletop amplified via a contact microphone. The seeds for the piece are two rhythms: the first, which I'll call A, is made up of two short beats followed by a long beat; the second, called B, is just a single long beat. Glass then instructs the player to combine the two units using a selection of regular arithmetic progressions. This is the algorithm that grows the seed.

The performer is given the freedom to choose their own algorithm, but Glass provides some examples of different arithmetical progressions that can be used to grow the piece. For example, AB AABBB AAABBBBB . . .

So the A rhythm increases by one each time, but the B increases by two. Much of the excitement for a listener is in solving the puzzle of how the piece is put together. As you listen, your brain begins to understand the sound isn't random, nor is it simple repetition, and you begin to crave an understanding of the underlying blueprint. Your ear is trying to reverse-engineer the construction. The music is tapping into the brain's addiction to structure as a way of making sense of the world. Although *1+1* is an incredibly simple example, its underlying challenge is perhaps a significant part of how we listen to music. Music is distinguished from noise by having some underlying pattern and logic at work which our brains recognise, a pattern that has been woven in by the composer either consciously or subconsciously. And it is the connection with patterns that aligns music so closely with the world of mathematics.

Einstein on the Beach was probably the culmination of this additive method that began with *1+1*. When you look at the scores for Glass's pieces based on this additive process, there are as many numbers as there are notes on the page, numbers that embody their overall algorithmic structure.

One can pick out two key moments in Glass's development as a composer which led to this particularly characteristic style of composition. Firstly, Glass actually studied mathematics as a student. At the age of 15, he entered an accelerated programme at the University of Chicago to study mathematics and philosophy. And it was during this time that he begun to understand the importance of aesthetics not just to the arts but to the composition of mathematics too.

As he said in an interview with the *Wall Street Journal*: 'Mathematicians are subject to the same kinds of enthusiasms as everybody else.' He was struck by how often mathematicians talked about beauty and elegance in their work. Aesthetics seemed to be as important as the concept of truth in the theorems they composed.

It was also at the University of Chicago that Glass first encountered the music of composers of the Second Viennese School: Schoenberg, Webern, Berg. This was music infused with very mathematical ideas, as I shall explain when I explore the blueprint of symmetry. But Glass felt that their ideas had already reached their sell-by date. Glass would joke that Arnold Schoenberg was the age of his grandfather. His music was no longer contemporary but belonged in the past. He was on the lookout for a new framework for his creative process. The second key inspiration for the development of his additive process was a meeting with the Indian musician Ravi Shankar.

Glass's epiphany came when Shankar asked him to transcribe a film score he had created full of ideas from Indian music for Western performers to be able to play. As Glass tried to write out the music, he found that using the Western convention of dividing music up into bars kept causing a strange effect. The beginning of a bar usually comes with an additional stress on the beat. But this wasn't in the original Indian music. It was like someone randomly stressing words in a sentence. Glass was becoming ever more frustrated with trying to fit the square peg of Indian music into this Western round hole.

When he mentioned his problems to the Indian tabla player Allah Rakha, he replied cryptically, 'All notes are equal.' Glass suddenly understood. Indian music didn't work on this vertical mindset of bar lines and harmony; it was a much more horizontal process, where the music grew and expanded as the piece evolved. It was this elastic quality of Indian rhythms that had attracted Messiaen too.

As Glass later wrote: 'What came to me as a revelation was the use of rhythm in developing an overall structure in music.' He liked to explain the difference between the character of western and Indian music by comparing the way western composers divided time to slicing up a loaf of bread where every slice was the same. In contrast Indian music took small units of time or 'beats' and laced them together to create music with much more rhythmic variety.

The Indian poets and musicians discovered how the Fibonacci numbers grow complex rhythms naturally from simple ingredients. For Glass too it is this additive procedure of growing more complex musical ideas out of simpler seeds that is central to Glass's sound world. For Le Corbusier, these numbers were key to the architecture for which he is famous, like his tower block in Marseilles.

Although the first reaction to this outwardly brutalist-looking building might be negative, for those who live inside the Unité, the structure's mathematics and its connection to natural growth create a space that they love. Many families, like that of Madame Gambu-Moreau, have passed their flats on to subsequent generations, who have continued to live happily there. As Jean-Marc Drut, who occupies apartment number 50, commented: 'The Unité d'Habitation is an incredibly powerful sculpture, and the feeling of living inside a work of art is a daily reality.' Each year since 2008 Drut invites an artist into his home to curate the design of his apartment, which he then opens for the public to experience.

But one of the curious features of the red and blue series, which are the blueprints for the building, is that they didn't have their origins in Le Corbusier's understanding of the power of Fibonacci numbers. Rather they came from his explorations of the geometry of the golden ratio. This important number first introduced by the ancient Greeks is the subject for our next blueprint.

Blueprint Four
The Golden Ratio

Novelists are always on the lookout for inspiration for their next novel. Eleanor Catton had been playing around with the Tarot as an interesting prompt for a new creative direction. Tarot cards have long been a catalyst for stories which fortune-tellers try to convince you might actually be your future rather than pure fiction. These cards were not, however, originally made for reading your fortune, but were part of a popular card game in Italy. The major arcana, consisting of cards like the Hanged Man and the Lovers, was simply meant to be a fifth, trump suit for a trick-playing game.

But the occult associations of the colourful range of characters used for this additional suit were a perfect vehicle for inspiring would-be fortune-tellers. As Catton played around with the cards she'd acquired, she was intrigued to discover that another author had used them in the creation of his novel. The Italian writer Italo Calvino had employed the Tarot as a curious device in his novel *The Castle of Crossed Destinies*. When Catton opened Calvino's book, she found an image of a strange layout of Tarot cards running vertically and horizontally across the page.

The conceit of the story is that the characters in the novel have travelled through a forest and arrive at a tavern where they mysteriously find they have lost the power of speech. Unable to tell their fellow travellers the story of their adventures, they instead use Tarot cards to narrate their journeys, which are then interpreted by the narrator. It is a novel that is meant to capture the challenge of trying to understand each other's consciousness through the fog of our limited means of communication.

Initially, Catton was very taken by this interesting blueprint for writing a story. But ultimately she felt that the novel failed.

Despite the fact that it is quite short, Catton found the book a real struggle. She was curious about why structure often led to books that somehow lacked life. Did the structure hinder readers enjoying the book rather than heightening the reading experience? She worried that structure got in the way of plot development rather than helping it. 'I thought about the novel that I wished *The Castle of Crossed Destinies* had been – and this, at last, was my negative-charge influence, defiant rather than imitative, longed-for rather than loved.'

The frustration at Calvino's failure gave rise to Catton's novel *The Luminaries*. It is a book that so successfully integrated a complex structural framework with an entertaining plot that it went on to win the Booker Prize in 2013. I actually love both books: the wonderfully vast, sprawling collection of characters that populates Catton's novel, but also the sparse, highly structured narrative of Calvino's. But then I guess I am a sucker for structure. Calvino has always been one of my all-time favourite authors.

One of the blueprints Catton used in *The Luminaries* came from the idea of employing horoscopes similar to the idea of fortune-telling that is at the heart of Calvino's story. She found a computer program online that plotted the planets as they moved through the signs of the zodiac. She dialled the program back to 1866, the year in which she wanted to set the story of her characters prospecting for gold in the New Zealand town of Hokitaka. Twelve of her characters represented the twelve signs of the zodiac, but her seven main protagonists corresponded to the sun, moon and five visible planets. The program allowed her to see how they moved relative to the stars in the sky.

As she tracked it over the year, she enjoyed the way that certain planets would follow each other through the night sky. What if the planets represented characters? The trajectories they mapped out might be an interesting structure for the paths that the characters took in the story. She found that Mercury, the planet that governs reason, was following just behind all the other players of the action. 'So I could build this narrative that the person who is trying to unravel the mysteries is one step behind it all.'

It wasn't just this astronomical blueprint that intrigued me, but another structural device that Catton employed to extraordinary effect. When you read *The Luminaries*, you find that each part is half the length of the previous one, and yet each part represents a month in the timeframe of the characters. The first part corresponds to January 1866 and occupies 360 pages. Once we get to the twelfth part, the narrative is served up in 16 lines. The effect is electric. Catton decided to implement a structure where each part contracted by the same factor so as to create a kind of spiralling pattern. The way the plot races towards the end is a direct result of the use of this mathematical tool to frame her narrative.

The mathematical factor Catton wanted to use was one that many artists have employed to create spiral structures: the golden ratio. The idea of this ratio, where A is to B as A+B is to A, felt like an interesting expression of the relationships between her characters. The book is all about prospecting for gold, so a ratio named after the valuable metal also seemed right. Catton intended the lengths of each part in relation to the ones next to it to conform to the golden ratio, but when she did the calculations she discovered that the result would have created an epic. Instead she scaled each part so that it was half the length of the previous one; even so, the novel is the longest to have won the Booker Prize. Nonetheless, Catton is part of a long line of artists who have been fascinated by the golden ratio as a blueprint for the imagination.

The golden ratio

As Eleanor Catton had learnt, the golden ratio is all about the relationship between two numbers. But it is also a geometric relationship. The sides of a rectangle are in the golden ratio if the ratio of the long side A to the short side B is the same as the ratio of the sum of the two sides A+B to the long side A.

This proportion was first identified as an interesting geometric idea in Euclid's *Elements*. There it is calculated as 1 to

$$\frac{1+\sqrt{5}}{2}$$

That is roughly 1 to 1.618; but like π and the square root of 2, this number has an infinite non-repeating decimal expansion. It is an irrational number, inexpressible as a simple fraction. It is traditionally represented by the Greek letter Φ.

These proportions crop up frequently across the natural and geometric worlds; a rectangle with them has lots of fascinating properties. One such might remind you of the property that the A-series of paper exploited, which I discussed in the previous blueprint. Instead of cutting the rectangle in half like the A-series, I am going to remove a square of dimensions B×B. A rectangle in the golden ratio has the special property that removing this square leaves behind a rectangle, whose dimensions are B by A−B, which is also in the golden ratio. It has the same proportions as the original rectangle.

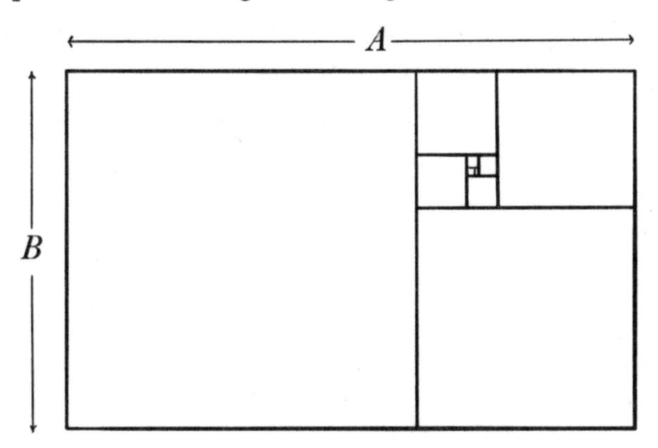

If I keep repeating the process, then I get a spiralling sequence of squares and rectangles that are all in the same golden ratio proportions. This picture might remind you of the one we built up in the previous blueprint using squares of Fibonacci dimensions. And there is indeed a connection. If I divide a Fibonacci number by its predecessor in the series, then the ratio gets closer and closer to the golden ratio the further through the series I go. This implies that a rectangle built from Fibonacci squares has proportions that get closer and closer to the golden ratio as the number of squares increases.

The Fibonacci spiral that we built from the bottom up is an approximation of the perfect spiral that emerges if you start with a rectangle in the golden ratio and work down. The spiral that you get from this process has the special property that the distance covered by it on each quarter turn increases by the same factor each time. This is an example of a 'self-similar spiral'. In the case of the golden rectangle, the length is multiplied by the golden ratio on each quarter turn. But you can create a faster or a slower spiral by choosing a different factor.

This spiral also goes by the name of a 'logarithmic spiral' or *spira mirabilis*. It was the seventeenth-century Swiss mathematician Jacob Bernoulli who gave the spiral the name 'miraculous' after he discovered this beautiful property of curves that retain their shape as they grow. He was so enamoured of the curve that he asked for the spiral to be carved on his memorial stone after his death with the inscription: *Eadem mutata resurgo* – 'Although changed I shall arise the same'. Although the stonemason got the epigram correct, unfortunately he carved the wrong sort of spiral into the monument.

These self-similar spirals are actually good examples of two other blueprints we will encounter: fractals and symmetry. A fractal structure has the property that if I zoom in or out, the shape remains the same – it is scaleless. These spirals can also be regarded as shapes with symmetry. I shall reveal later how objects have symmetry if there is some action you can do which transforms their shape but preserves their structure. Often that's about reflecting or rotating them; in the case of the spiral, it's about scaling the shape. Magnifying or shrinking preserves the structure. It is an example of scaling symmetry.

The importance of these self-similar shapes extends beyond mathematics. They occur all across the natural world. Cyclones often develop with spirals demonstrating this self-similarity. These spirals are the blueprints for galaxies like our own Milky Way. Hawks will hunt their prey by following a self-similar spiral. As the prey flees in a straight line, the spiral ensures the view of the prey is always projected onto the most sensitive part of the hawk's retina as it gradually converges on the victim. Shells of molluscs grow in this self-similar spiral. As the mollusc inside the shell grows, it needs to enlarge its home. The mollusc keeps things simple by just multiplying the dimensions of the previous chamber it inhabited in the shell by the same factor each time.

The most famous example of this is the nautilus shell. This beautiful orange-brown tiger-striped shell is the poster child of every website, magazine article or book that wants to excite the reader about mathematics underpinning the natural world, especially when it comes to the existence of the golden ratio or Fibonacci numbers in nature. But despite having seen the shell so many times, I must admit that I got rather a shock when I came face to face with a live nautilus for the first time.

They are bizarre, alien-looking creatures. Two large eyes stare out at you from their shell, and they have strange tentacles that they use to explore their environment. They are naturally active at night and they like darkness, living in deep water. They are essentially an octopus with a shell. Their lineage is hundreds of millions of years old, and they really haven't changed very much in all that time.

I was quite intrigued by the question of what the scaling factor was for the nautilus's beautiful shell. So I got hold of a shell from a friendly lab and started measuring. Here are the distances from the centre of the spiral shell to each of the different chambers that the mollusc had built:

3.07, 3.32, 3.59, 3.88, 4.19, 4.52, 4.88, 5.27.

If I take 3.32 and divide by 3.07, I get 1.08. Divide 3.59 by 3.32, and I get 1.08. Take 3.88 and divide by 3.59, and I again get 1.08. Every time I did this calculation, I got the same number. When the nautilus builds a new room, the dimensions of that room are 1.08 times the dimensions of the previous one. It's by following this simple mathematical rule that the nautilus builds its elegant spiral. Now, the nautilus builds on average four chambers for each quarter turn, which means the scaling factor for the nautilus spiral is $1.08^4=1.36$. This is well short of the golden ratio of 1.618.

So, a big warning here. Most self-similar spirals have nothing to do with the golden ratio. You can generate a spiral using any number as a factor to determine how much its length increases at each quarter turn. So many people have perpetuated the myth that the nautilus shell is a golden spiral encoding the Fibonacci numbers that it's even found its way into Dan Brown's *The Da Vinci Code*.

Weirdly, though, my nautilus's scaling factor of 1.36 is very close to the square root of the golden ratio, 1.355674 . . . A statistical analysis by Christopher Bartlett in 2018 of 15 specimens of the nautilus species *Allonautilus scrobiculatus*, otherwise known as the crusty nautilus, found that the mean scaling ratio was 1.356. That is so spookily close to the square root of the golden ratio as to have sparked a whole new myth that the nautilus is actually tapping into the golden ratio after all. But the truth is that most spirals in nature are not golden.

The mathematical gardener

But there is something rather special about the golden ratio spiral in nature which is different from any other spiral and which is actually going to reveal why we see the Fibonacci numbers all over the natural world.

The number of petals you see on a flower is the end result of a process that starts at a micro-scale. As the new shoot of a plant grows, plant tissue called 'primordia' are added to its tip, gradually building the body of the flower outwards. One can quite clearly see the structure of this cellular growth replicated at larger scale in, for example, the seedhead of a sunflower.

The interesting question for a plant, having built its structure so far, is where the best place is to grow the next primordium. It turns out that the answer comes down to finding the most efficient way to pack these new cells together as they grow. They want to be far from the last cells that were grown, to avoid competition for light and to maximise their exposure to the sun. In 1837, the French physicist Auguste Bravais – who would become famous for his work on crystal lattices – observed with his brother Louis that leaves on the stem of a plant would grow at an angle of 137.5 degrees from the previous leaf that had appeared on the same stem. This arrangement helped to maximise the leaves' contact with sunlight. Once plants could be analysed on a sufficiently microscopic scale, the same angle was discovered to determine the addition of each new primordium.

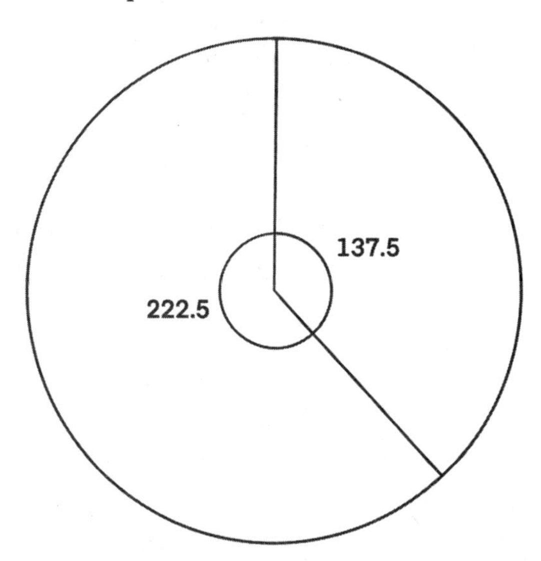

This angle is often known as the 'golden angle'. This is because if you measure the angle around the outside the circle, rather than internally, then it is 222.5 degrees. If you calculate the ratio of a full circle of 360

degrees to this angle determining the growth of plants, you get 360/222.5=1.618. The golden ratio. The golden angle is the angle which most efficiently packs new cells into the shoot as it grows while keeping them away from the most recently grown ones.

This efficient packing is actually related to a strange property of the golden ratio. It is the most irrational of all irrational numbers. What does this mean? It's not an easy answer but if you take an irrational number X, then for each number N there will be a fraction a_N/N that gets closest to X for that fixed denominator N. As you increase N, you can ask how well does the fraction a_N/N approximate X. The golden ratio Φ is the irrational number that is slowest to be approximated by fractions. Irrationality is key to growth because if you laid cells down using fractional divisions of the circle, then you'd just get radial lines and the plant would collapse very quickly. So an angle that is as far away from a fraction as possible is optimal for creating the densest packing of cells as the plant grows.

Computer simulations illustrate why a flower collapses very quickly if you deviate from this golden angle even slightly: there is too much unfilled space in the structure to hold the plant together.

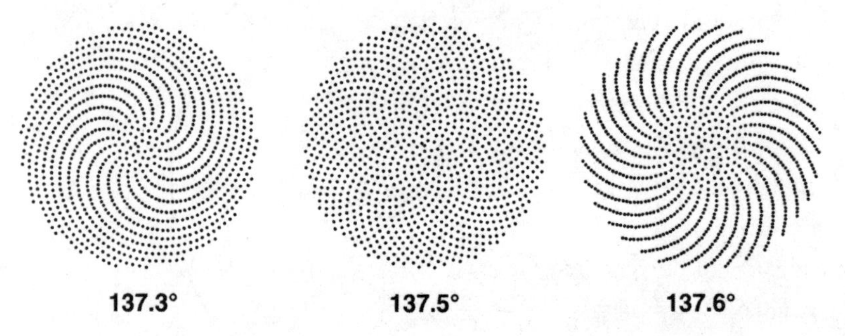

The reason that you see Fibonacci numbers of petals in a fully mature flower is because the ratio of two successive Fibonacci numbers is a good approximation to the golden ratio that has grown that plant. As a plant grows, sometimes you can see the spirals – like the ones in the head of a sunflower or on the outside of a pineapple – that encode successive Fibonacci numbers. The spirals get more complex as the plant develops. But whenever the plant decides it's time to grow petals or leaves, this will happen at the ends of these spirals. You can't have an irrational number of petals, and so you see the whole number approximations from the Fibonacci sequence appearing.

The speed at which a plant grows is also a key factor to determine which Fibonacci numbers appear. As a stem continues to grow upwards, older primordia are moved outwards and downwards from the tip. The rate at which this happens plays a role in the way a plant matures. In experiments carried out in the 1990s, the physicists Stéphane Douady and Yves Couder used magnetic droplets of liquid in place of plant tissue, and they let them fall into a dish filled with silicone oil. The edge of this dish was magnetised so that the droplets were attracted to it as well as being repelled by each other. This simulated the idea that new cells sought positions that were far from cells that had grown just before them. In plants, this repulsion between successive primordia is achieved through the production of a hormone called 'auxin', which creates an effect similar to the magnets.

The drip rate, which simulated the rate of growth of the primordia, turned out to be a key factor in the way these droplets would form into golden spirals. Faster or slower rates created different Fibonacci numbers. But the experiments also revealed an intriguing coda to this story. There is a second, slightly less efficient angle of 99.5 degrees, which also seems to successfully grow a flower. If you analyse the number of petals that emerges from this process, you get the following alternative Fibonacci sequence: 1, 3, 4, 7, 11, 18, 29, . . . So this might explain those few exceptional flowers, like the starflower mentioned in the last blueprint, that have seven petals.

So although the five petals of the pansy or the 13 petals of the marigold are genetically programmed into the DNA of these plants in some manner, we now understand that it is a secondary result of a more fundamental bit of coding which uses the golden angle to determine growth. The maths comes first, then the biology. The computer simulations show that a plant that isn't tapping into the efficiency that this golden angle provides will not grow very well and will quickly fail to replicate itself. The golden ratio is the number that survives, and it results in flowers that favour Fibonacci. But the golden ratio is not just key to nature's creations.

The cult of the golden ratio

For Le Corbusier, the golden ratio was a proportion that naturally appeared from his investigations of the square as a basic building block in architecture. Take a square and draw a line from the midpoint of one side to the opposite corner. Keeping the midpoint fixed, swing this line round so that it aligns with the side of the square. Use this extended line to build a rectangle out of the square. This rectangle will be in the proportions of the golden ratio. It was this discovery which sparked Le Corbusier's journey that culminated in the Modulor Man becoming the blueprint for his buildings.

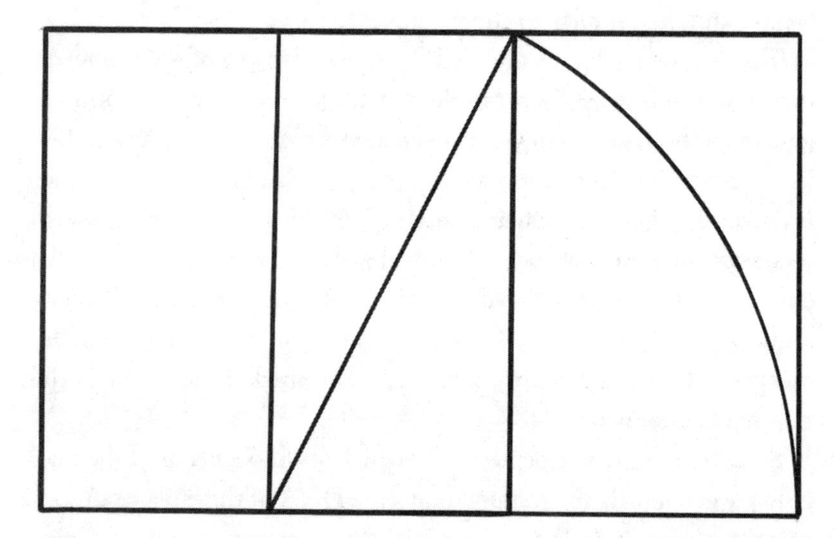

As I discussed in my blueprint dedicated to the circle, the ancient Greeks were particularly interested in geometry and the challenge of creating geometric shapes using only the simple tools of a compass and a straight edge. It was this that led them very quickly to the golden ratio, since the proportion often emerges in shapes made with this basic toolkit. For example, the construction of a pentagon detailed in Euclid's *Elements* reveals that the golden ratio is intimately related to many of its internal proportions. Construct a five-pointed star inside the pentagon by drawing lines from each vertex to the two opposite vertices and these lines are in the golden ratio to the sides of the pentagon.

This is the reason that the five-pointed star is such a ubiquitous symbol in the culture of the Freemasons. It has nothing to do with strange devil worship or witchcraft but rather goes back to the fundamental origins of the Masonic order: they were builders. The straight edge and compass were basic tools of the trade for a mason. Euclid's *Elements* is like a bible for the order and actually plays a significant role in many of its rituals. The five-pointed star is in fact a measuring device like a set square for the golden ratio and isn't related to magic and mysticism.

The proportion received a second burst of interest in the Renaissance, and it was during this period that art and mathematics in Europe became intimately related. The fifteenth-century Italian mathematician Luca Pacioli is probably more responsible than anyone for the elevation of the golden ratio to an almost mystical blueprint for art, nature and the universe itself. Pacioli is most famous for his development of double-entry book-keeping for accountancy. But in 1509 he published a three-volume treatise on the importance of the golden ratio – or what the title of his work was first to call the *divina proportione*, the 'divine proportion'. The book recommended itself as:

> A work necessary for all clear-sighted and inquiring human minds, in which everyone who loves to study philosophy, perspective, painting, sculpture, architecture, music and other mathematical disciplines will find delicate, subtle and admirable teaching and will delight in diverse matters touching upon a very secret science.

From this bombastic opening, Pacioli went on to philosophise about the mystical and divine importance of the ratio to the very fabric of the universe. It is God's blueprint for creation. Its irrationality is related to the unknowability of God. The three numbers that define the ratio are compared to the Holy Trinity. The five Platonic solids, which are connected to the five classical elements from which everything is made, Pacioli explained, are infused with this proportion.

But it was the artist he enlisted to represent these three-dimensional shapes on the two-dimensional page that helped propel his book into the classic it is regarded today. Leonardo da Vinci supplied a beautiful series of skeletal images that helped bring the mathematics alive and allow people to see for the first time some of the geometric shapes mentioned in the book.

Leonardo's connection with Pacioli and the *divina proportione* is probably responsible for fuelling the huge industry that has emerged to try to prove that Leonardo used the golden ratio as a blueprint for his art, from the *Mona Lisa* to the *Virgin of the Rocks*. There is certainly evidence that he was interested in mathematical proportions in the human form. His sketchbooks include a man in profile with a grid of squares and rectangles overlaying his face. You also find in his notebooks his musings on mathematics: 'He who does not know the supreme certainty of mathematics is wallowing in confusion.'

The golden ratio is probably the most ubiquitous number to crop up in the dialogue between the world of art and the world of mathematics, often spuriously. Some artists have embedded golden ratios in their canvases knowingly, like Salvador Dalí. Others seem to have been drawn to the ratio more intuitively. For example, Piet Mondrian's paintings are famous for their arrangement of coloured rectangles across the canvas, and research has revealed a large preponderance of those rectangles with proportions close to the golden ratio.

But works from Raphael's *Madonna of the Goldfinch* to Rembrandt's *The Anatomy Lesson of Dr. Nicolaes Tulp* have also been offered up as examples of the golden ratio in painting. There is, it has to be admitted, something of a cult for finding golden ratios in the work of artists, and it is important to be a little sceptical about all these occurrences.

But there has been some research that offers a possible reason why our eyes are attracted more to canvases that encode this ratio than other ones. Adrian Bejan, professor of mechanical engineering at Duke University, in Durham, North Carolina, has found that the human eye is much more efficient at accessing information that is encoded in an image in the golden ratio. There seems to have been an evolutionary advantage to being able to scan quickly a region that has these dimensions. As we looked across the savannah for animals that we could hunt, or that might hunt us, the lie of the land to be scanned was longer on the horizontal than the vertical. The optimal area for the brain to be able to scan, Bejan has discovered, is in the golden ratio.

It seems that shapes in these dimensions are easier for the eye to scan and to communicate information to the brain. Bejan writes in his paper how 'Animals are wired to feel better and better when they are helped and so they feel pleasure when they find food or shelter or a mate.' This is why he believes that we then associate the idea of the golden ratio with beauty because anything that helps an animal to navigate its way through the world is going to make it experience gratification.

This ability to scan quickly might in part be due to the natural spiral that sits inside the rectangle in the golden ratio. Remember how the hawk uses this spiral to keep its prey optimally in sight. It's interesting that an artist such as Leonardo does seem to position interesting things in a canvas at the points where these spirals direct the eye, rather than simply putting them symmetrically in the middle of the canvas.

Mathematical ideas were always important to Leonardo. His famous sketch of *Vitruvian Man* is an exploration of the relationship of mathematical shapes to the human body. In fact, what most people don't realise about this iconic image is that it is a solution to a conundrum that had baffled people for 1,500 years. The Roman architect Vitruvius wrote about the importance of proportions in the human body to the creation of buildings in his seminal work *On Architecture*. In one passage, he talked about the mathematical relationship of geometric shapes to those proportions:

> The navel is naturally placed in the centre of the human body, and, if in a man lying with his face upward, and his hands and feet extended, from his navel as the centre, a circle be described, it will touch his fingers and toes. It is not alone by a circle, that the human body is thus circumscribed, as may be seen by placing it within a square. For measuring from the feet to the crown of the head, and then across the arms fully extended, we find the latter measure equal to the former; so that lines at right angles to each other, enclosing the figure, will form a square.

The challenge: how to place the image of a man simultaneously inside a square and inside a circle with its centre at the man's navel. The square picks up on the interesting observation that a person's height is often the same as the span of their outstretched arms. Before Leonardo's diagram, people always fell into the trap of aligning the centre of the circle and centre of the square. But when a man was placed inside this arrangement, the proportions were always wrong. It was Leonardo's inspired move to disconnect the centres of the shapes and shift the square down that solved this problem. This asymmetry produced two geometric shapes that housed the perfectly proportioned man.

I had the chance while on a visit to the Accademia in Venice to get access to the original diagram that Leonardo drew. What is fascinating is to see the page punctured by little holes that obviously correspond to the compass that Leonardo must have used to lay out the diagram with mathematical precision.

For Vitruvius, the mathematics was a way to translate the proportions of the body into those of the buildings they would occupy. And by using the mathematical shapes as blueprints for the building, you would create spaces that were organically in tune with the human body. Every architect since Vitruvius has grown up with this idea of the deep connection between building and body, with mathematics as the mediator between. It was extending these ideas that led Le Corbusier to propose his new geometric template for the architecture of the twentieth century.

The Modulor

In the early 1920s, Le Corbusier had caused something of a storm by writing about his ideas for the mass production of buildings and for rebuilding the centre of Paris. He described houses as 'machines for living in' and asserted that 'mass production, machine, efficiency, cost price, speed, all these concepts called for the presence and the discipline of a system of measuring'. It was his drive to create such a system that led Le Corbusier to the idea of the Modulor. As we saw in the previous blueprint, music was an important inspiration for him. His Modular was like a musical scale that could be used to make something beautiful or simply banal. Le Corbusier's Modulor was to be a scale for space rather than sound. But inside the name he gave this scale is the secret to its construction. The name combines the concept of the 'module' with that of the *nombre d'or*, the French term for the golden ratio. It was this ratio that was the catalyst for Le Corbusier's experiments in creating a system of measuring.

The French standards organisation was proposing a system of building construction that was based on simple arithmetic, but Le Corbusier considered it arbitrary and not fit for purpose. He believed that any system of measurement for architecture had to also apply to the way things grow in nature. 'Take trees: if I look at their trunks and branches, their leaves and veins, I know that the laws of growth and interchangeability can and should be something subtler and richer.' The secret he believed was to be found in mathematics. This was the language that could provide the link between all these forms of growth.

Le Corbusier became convinced that his diagram of the extended square which produced the golden ratio had much more to give. His notebooks began to fill with experiments depicting lines and arcs and right angles growing out of this golden ratio rectangle. But he did not lose sight in these diagrams of the importance of reflecting the proportions of the human body. He saw himself not just as a modern-day Pythagoras but as a contemporary Vitruvius.

His central observation was that two squares on top of each other will perfectly house a man with arm aloft with the midpoint of the structure centred on the man's navel and his feet astride the base of the square. His experiments were an attempt to connect this idea to the diagram producing the golden ratio. He believed that a third square, relating to the golden ratio, could naturally be placed inside this frame, leading to dimensions that would reveal the hidden mathematics of the man standing inside with his arm raised. The assistants in his studio were tasked with solving this modern-day version of Vitruvius' conundrum.

By 1943, they had come up with a complex geometric design they called 'the Grid', which appeared to solve the challenge. During a stormy crossing of the Atlantic to the USA on board a cargo ship in December 1945, Le Corbusier was allowed to work in the cabin of one of the officers for eight hours every day while the man was on duty. 'I am not going to leave this confounded boat before I have found the explanation of my golden rule.' The diagrams he created on board led naturally to a scaling of the Grid by factors of the golden ratio. And it was from there that the dimensions corresponding to the red series and blue series emerged. The red corresponded to the scaling applied to the single square, the blue to the height of a man standing inside the double square with arms aloft. This was where the wavy lines that accompanied Modulor Man made their first appearance.

One of the issues Le Corbusier had in creating his Modulor was that applying the golden ratio to the heights of the squares produced a series of irrational numbers which would be completely impractical for a builder to use as dimensions. And so he was forced to round up or down the measurements that arose from his calculations. That is why the numbers in the series don't quite match the perfect Fibonacci rule.

Le Corbusier also wrestled for some time with what the height of his squares should be. His trip to America took him from the metric world to the world of feet and inches. He became convinced that his Modulor would become an essential tool in navigating the pitfalls of swapping between these two systems of measurement.

Le Corbusier's original proposal for the height of the man had been 1.75 metres. 'But isn't that rather a *French* height?' pointed out one of Le Corbusier's team. 'Have you never noticed that in English detective novels, the good-looking men, such as the policemen, are always six feet tall?' That's eight centimetres taller than the French height. When Le Corbusier switched the Grid to accommodate this handsome policeman measured in feet and inches, he found that the translation from metric into imperial measurement seemed to magically produce a new Modulor made up of round figures.

By now Le Corbusier felt confident that he had discovered a new system of measurement and he went to log his achievement at the patent office in Paris. Le Corbusier tried to explain his idea twice, but the patent officer remained baffled: 'I'm afraid I don't understand . . .' But Le Corbusier was persistent, and by the third explanation the officer began to soften: 'Hold on a minute, I am beginning to see, this seems to be something extraordinarily interesting.' It's probably more the case that the officer was bullied into submission.

On one of his trips to the USA, Le Corbusier met with Einstein at Princeton University. He was very excited to see what the great tamer of the geometry of space-time would think of his discoveries.

But as he tried to explain his ideas on the Modulor he found himself getting mired in details, unable to articulate clearly the discovery he'd made. He felt he needed to talk in the language of mathematics to the great scientist and was excited when Einstein picked up a pencil and started to explore the properties of Le Corbusier's diagrams. 'Stupidly I interrupted him, the conversation turned to other things, the calculation remained unfinished.'

Einstein very graciously sent Le Corbusier a note that evening: 'It is a scale of proportions which makes the bad difficult and the good easy.' My impression is that Einstein was fighting to find something encouraging to offer the architect. The truth is that, like the patent officer and Einstein, I too have found the diagrams that Le Corbusier was producing rather baffling. And after some calculations, I realised that Le Corbusier was just wrong. His geometry was faulty.

His diagrams seem to show the three squares naturally arising from his musings over the golden ratio rectangle. But the trouble is that the geometry implies the two squares in which he placed his Modulor Man are not actually true squares. They are necessarily 0.6 per cent longer in one direction than the other. On the page, this discrepancy is not apparent. They look square. But a 0.6 per cent discrepancy in creating a building of the dimensions that Le Corbusier was proposing would be devastating.

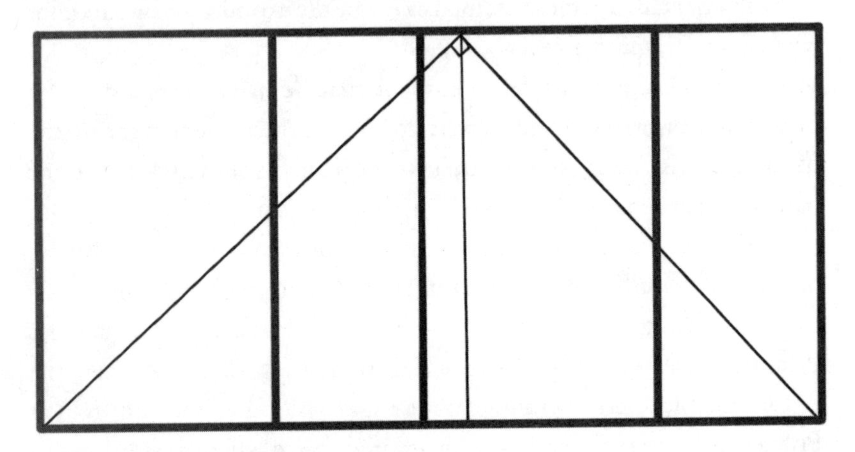

Le Corbusier believed that mathematics was the blueprint of the universe itself and felt impelled to frame his ideas in the language of mathematics. Even when René Taton, a mathematician at the Sorbonne, took the time to reveal Le Corbusier's errors to him, Le Corbusier still felt that these were minor quibbles. He did at least publish Taton's mathematical critique in his notes on the Modulor, but it is clear that he really just wanted to be able to quote a respected mathematical source in support of his ideas.

It took the arrival at Le Corbusier's practice of a young Polish architect, Jerzy Sołtan, to declare the emperor had no clothes and have the guts to challenge the mathematical thinking of the great architect. 'Corbu was not strong in mathematics, but he was very much under its spell,' Sołtan said. After several days wrestling with the ideas behind the Grid and the construction of the red and blue series, Sołtan had a strong reaction against Le Corbusier's claim to have made a new discovery:

> It seems to me that your invention is not based on a two-dimensional phenomenon but on a linear one. Your 'Grid' is merely a fragment of a linear system, a series of golden sections moving towards zero on the one side and towards infinity on the other.

'All right,' Le Corbusier replied, 'let us call it henceforth a *rule* of proportions.'

But Sołtan was right. Essentially Le Corbusier had just invented again a version of the Fibonacci sequence. And even then he had slightly messed things up by having to approximate the golden ratio. You can hear his frustration when eventually he declared: 'To hell with the Modulor. When it doesn't work, you shouldn't use it.'

Divine mathematical music

Proportions are important in visual art and architecture, but the golden ratio and the Fibonacci numbers have also played a significant role as a blueprint for musical composition. We saw how the Fibonacci numbers had their origins in the rhythms of India, but the natural sense of growth that these numbers encode has also been of interest to Western composers. In his *Music for Strings, Percussion and Celesta*, Bartók used the golden ratio and Fibonacci numbers to mark significant moments throughout the piece. Here is a diagram highlighting those key moments in the first movement.

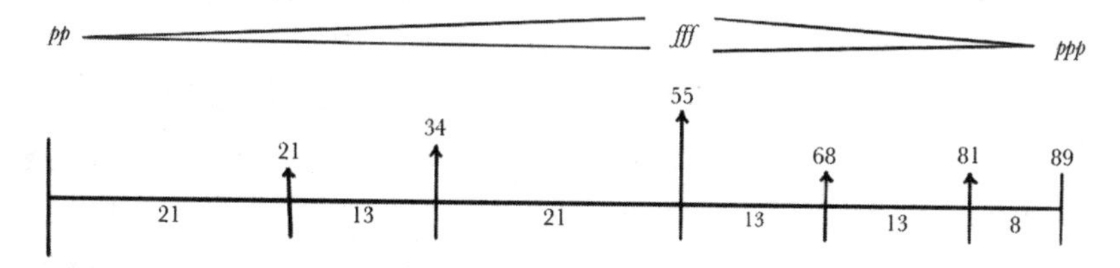

By constructing the piece out of Fibonacci numbers, Bartók created an overall structure which reflects the golden ratio. Even the harmonic landscape is built out of the golden ratio and the Fibonacci numbers.

In classical music, the scales which had been used up to this point had a very additive character. They are based on the different wavelengths that fit a string, and are in a 1 to 2 to 3 to 4 to 5 ratio. The series of notes you can generate from this are called 'overtones'.

Bartók decided to experiment with a scale that was multiplicative in nature such that the frequency of each note in the scale is a fixed multiple of its predecessor. Much like Eleanor Catton's original plan for the structure of *The Luminaries*, each note's frequency would be in the golden ratio to that of the next note, and Bartók used the Fibonacci numbers to do this. His new scale was built from notes in proportions 8 to 5 to 3 to 2. The result combined the qualities of both a major and a minor scale. Bartók believed this was more natural to our ear, and it is very much associated with folk music. The scale's multiplicative character grows like a spiral, and some believe that the spiral nature of the inner ear – the logarithmic structure of the cochlea – means that we are more readily responsive to a scale based on Fibonacci numbers.

Bartók once wrote: 'We follow Nature in composition.' Given that the golden ratio and the Fibonacci numbers appear all over nature, it's not surprising that we find Bartók drawn to these natural numbers. Mathematics is the language of nature. Perhaps this is why we hear composers exploring mathematical themes in their music. They are simply responding to nature's language. Sometimes this is an intuitive attraction, but in others, as with Bartók, it is a conscious tapping into the mathematician's cabinet of wonders.

Claude Debussy too was drawn to the natural generative power of these numbers in his compositions, from large scale works like *La Mer* to his more intimate pieces for solo piano. While Bartók talked openly about his use of Fibonacci numbers in his music, Debussy was less forthcoming. He hated the way music was analysed by theorists and its magic laid bare: 'Grownups . . . still try to explain things, dismantle them and quite heartlessly kill all their mystery.' No wonder, then, that he didn't divulge the blueprints for his music. It was the detective work of musicologist Roy Howat in uncovering the golden ratio and Fibonacci numbers at key points in Debussy's compositions which convinced scholars that the composer was very conscious of what he was doing.

Howat supported his argument with the fact that numerical structure was extremely important to French poets of the time. As Baudelaire wrote in his work *Fusées*: '*Tout* est nombre. Le nombre est dans *tout*. Le nombre est dans l'individu. L'ivresse est un nombre' – '*Everything* is number. Number is in *everything*. Number is in the individual. Intoxication is a number.' Debussy admired Baudelaire's poetry and set a number of his poems to music.

Howat's amazing analysis of Debussy's masterpiece *La Mer* highlights how its structure is intimately related to the Fibonacci numbers and the golden ratio. Here is one of Howat's diagrams analysing the third movement of the piece, the 'Dialogue du vent et de la mer' or 'Dialogue of the wind and the sea'.

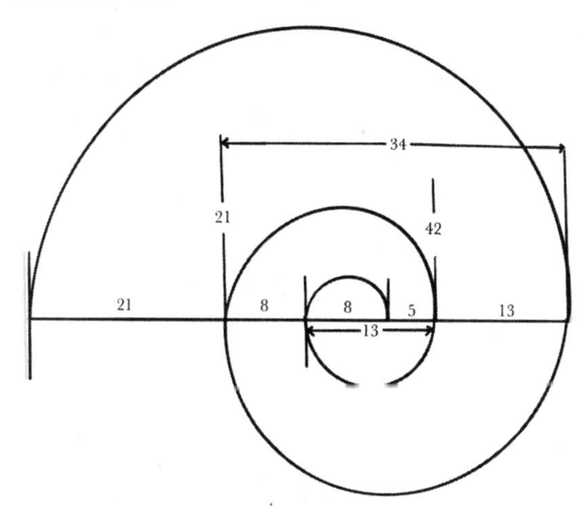

What is fascinating is that, rather than employing the Fibonacci numbers in their strict sequence, as Bartók did, Debussy mixed them up. But as Howat revealed, his choices were not random but controlled by a spiral that they made in the music, as if he was trying to create in sound the vortices of water that a storm whips up at sea.

That golden ratios and Fibonacci numbers were deliberately at work in Debussy's piece can perhaps be further confirmed by another decision that the composer made. The first edition of the score of *La Mer* appeared with a reproduction of Katsushika Hokusai's famous woodblock print *The Great Wave off Kanagawa* on its cover. It is an image that Debussy had hanging on the wall of his study. There is a photograph of the composer with Igor Stravinsky where you can see the framed print in the background. I will revisit Hokusai's work in the blueprint on fractals. But there is a reason why Debussy put it on the cover beyond it just being a beautiful depiction of the sea – *la mer*. Hokusai's picture, as well as being an example of a fractal, is proportioned in the golden ratio. If you measure key points in the painting, such as the peak and the trough of the wave and compare them to the overall dimensions of the work, there are golden ratios everywhere.

As well as his large scale compositions, Debussy also proportioned some of his piano pieces using this ratio. And it was here that Howat discovered the first real hint that the composer was deliberately doing this. In August 1903, he wrote an interesting letter to his publisher which perhaps betrayed his conscious use of the golden ratio in his work. In the letter, Debussy mentioned that he had accidentally missed out a bar from the original score of his piano piece 'Jardins sous la pluie'. When you see the bar he added, you'll wonder what all the fuss is about; it simply repeats the previous one. But then he explained why it should be included: 'it's necessary, as regards number; the divine number, as Plato would say'. The 'divine number' is a reference to the golden ratio. Without this extra bar, the mathematical proportions of the piece don't work.

It transpires, though, that Bartók and Debussy were not the first classical composers to embed golden ratios in their music. There is evidence of Mozart doing the same thing in *The Magic Flute*.

Composing by number

I love Mozart's operas, so when the Royal Opera House asked if I'd like to take over its Linbury Theatre for a run of performances to explore connections between mathematics and music I couldn't resist using one of them as a blueprint. I spent two glorious weeks in rehearsal with five singers and a *répétiteur*, putting together an interactive exploration – with lots of audience participation – of the way Mozart used maths in his music.

Of all Mozart's operas it is the last one, *The Magic Flute*, which is most laden with symbolism and numerical imagery. So I decided to take the audience in the Linbury Theatre on a crazy mathematical journey through this beautiful opera. I think what we created is one of my proudest attempts at showing mathematics to be a blueprint for the arts. The reason *The Magic Flute* is dripping with maths is partly due to the addition of a very mathematical strand to Mozart's life seven years before the opera's completion: his initiation into the Masonic order in 1784.

In contrast to the modern image of the Freemasons as a club for the establishment, in the late eighteenth century the order was very much driven by the ideals of the Enlightenment and the revolutionary changes sweeping Europe. It was considered such a threat to the authorities that the Austrian empress Maria Theresa had banned the order twenty years earlier. In fact, many have suggested that the Queen of the Night was modelled on the empress, while Sarastro's order is really a thinly disguised version of the Masons.

With *The Magic Flute*, Mozart was pinning his colours to the cause. He did this by the use of the number 3, which is hugely significant to the Freemasons. A Masonic song, for example, includes the lyrics:

> *So drink my brothers drink*
> *Three times three as true masons.*

Mozart as a new initiate would have knocked out a rhythm of three on the door of the lodge to gain entry. And over and over again in the opera, starting with the overture, we hear this rhythm of three, the triple chord, often repeated three times. As Goethe, a fellow Mason, pointed out: 'the crowd should find pleasure in seeing the spectacle: at the same time, its high significance will not escape the initiates'.

Why the number 3 is so important to the Masonic order is somewhat shrouded in mystery. Some have picked up on the importance of 3 to the Pythagoreans, who regarded it as the first male number. Others have speculated there was a more Christian connection, although the Masons didn't associate themselves exclusively with one faith. But the most interesting explanation of the importance of the number 3 relates to the Masonic connection with geometry. After all, the Freemasons evolved from medieval guilds of builders. The oldest Masonic text begins with the history of the craft of masonry and describes how it all originated with Euclid, the inventor of geometry. One of the rituals of the Masons relates to the 47th postulate of Euclid, which is all to do with Pythagoras' theorem about the right-angled triangle, a shape used by any mason. The significance of 3, therefore, might relate to the importance of the triangle to the builders' craft.

The three chords that pepper the score of *The Magic Flute* aren't the opera's only use of 3 to denote this Masonic undercurrent. Lots of things come in threes. Tamino is rescued at the beginning of the opera by three ladies. He is guided to Sarastro's temple by three boys, or genies. When he arrives, he finds three temples, dedicated to 'Nature', 'Reason' and 'Wisdom' – three Masonic ideals. Mozart often wrote in E flat major, a key with three flats, if he was composing music with a Masonic theme. Trios abound, giving rise to three-part harmony. Chords based on the third and sixth note of the scale play an unusual part in the musical composition.

If 3 is Mozart's leitmotiv for the Masonic themes at the heart of the opera, then the number 2 is employed to indicate a society in transition, full of unions and oppositions. In contrast to the male association of the number 3, the number 2 has in many cultures been considered a female number. The opera is full of significant pairings. Tamino and Pamina. Papageno and Papagena. Moon and sun. Earth and air. Fire and water. The glockenspiel and the flute. Not all these pairings are positive. The Queen of the Night stands in opposition to and in conflict with Sarastro. In our interactive exploration at the Royal Opera House, as Pamina and Papageno sang the duet 'Bei Männern', members of the audience had to choose which side of the pairings they most identified with. Are you night or day? Fire or water?

One of the central thematic pairings in the opera is the tension between chaos and order, the movement from an old politics based on the rule of kings and queens to new liberal principles, from religious superstition to rationalistic enlightenment. This is Mozart writing at a crucial moment in history, both politically and musically. The work premiered in Vienna in 1791, as revolution swept through the streets of Paris. The Masons were at the time playing an important role in that transformation. One of the sayings of the Freemasons declares: *Ordo ab Chao* – 'Order out of Chaos'. This passage from chaos to order is represented in Masonic temples by a raw block of stone in their northern corner and a polished perfect cube in the southern.

But the opera is also a statement of a musical revolution. The music of the Queen of the Night is full of coloratura; it is ornate, Baroque. But the classical music of Sarastro's temple, and the music that accompanies the trials that Tamino and Pamina undergo is much sparser, more complex harmonically if simpler in musical line.

As our Queen of the Night wandered the stage singing her famous aria 'Der Hölle Rache', we got our audience to play a theatre game that simulated chaos in action. Each person had to move around the auditorium so as to create an equilateral triangle with two other audience members that they'd secretly chosen. The trouble was that those two would be trying to form equilateral triangles with other people. It is a very simple algorithm, but the result of these interlocking triangles is chaos. In one performance, a member of the audience shrieked with fright when the Queen of the Night started to sing just behind her. One isn't usually so close to the powerful voice of an opera singer.

The overture that begins the opera is a microcosm of this transition from the chaos of the music of the Baroque to the new musical order that Mozart introduces. The opening consists of a lot of offbeat rhythms and destabilising sounds which eventually resolve into something more stable.

But the most significant piece of mathematics that I suspect Mozart learnt from the Masons is hidden inside the mathematical symbol which has come to represent the Masonic order and which can be seen carved into the pavement outside the grand Freemasons' Hall in Covent Garden: the five-pointed star, or pentagram, and the golden ratio encoded inside its geometry. The number 5 is the fusion of 2 and 3, the female and male, a theme very much at the heart of *The Magic Flute*. The future success of Sarastro's order depends on the union of man and woman. It is striking that, in the opera, duets and trios give way not to quartets but to quintets.

For me it is the way that Mozart musically manifests the golden ratio within the pentagram that is most extraordinary. If you count the number of bars from the start of the first allegro in the overture to the point at which the important triple chords are heard, you get 81 bars. Now count from the triple chords to the end of the overture. You get 130 bars. Divide 130 by 81 and it is very close to 130+81 divided by 130. This is the closest you can get to the golden ratio with this number of bars. Was Mozart intuitively drawn to this mathematically important concept, or did he deliberately design his overture this way? Mozart's time as a Freemason must have exposed him to the idea of the golden ratio, so it's hard to believe it wasn't planned.

Mozart also seems to have used the Fibonacci numbers to repeat the same structure on a much greater scale. Is it just a coincidence that he divided the two acts of *The Magic Flute* so that the first comprised 8 musical sections and the second 13, making 21 in total – all three of which are Fibonacci numbers? This has the effect of dividing the whole opera into the golden ratio. The microcosm of the overture is reflected in the overall shape of the entire work.

Mathematics was important to Mozart even before he joined the Masons. There is a lot of evidence that he had fallen in love with numbers from a very early age. Johann Andreas Schachtner, Salzburg's court trumpeter and a friend of the Mozart family, wrote about the young Wolfgang's extraordinary appetite for figures: 'when he was doing sums, the table, the chair, the walls and even the floor would be covered with chalked numbers'.

Mozart's obsession with numbers didn't wane. As an adult, he would scatter numbers throughout the letters he wrote to family and friends. His family used a secret code in order to keep politically sensitive comments from the eyes of the censors. But Mozart also used numbers in more intimate exchanges. His kisses would invariably be issued in units of a thousand, although sometimes he would choose more interesting numbers to shower his correspondent with.

One of the letters Mozart wrote to his wife Constanze contains a curious string of numbers: 1095060437082. One decoding of this sequence suggests adding the numbers 10+9+50+60+43+70+82 to get 324, which is 18^2. For Mozart, the idea of a square number expressed the bond of love between him and his wife. He signed off a letter to his sister as a 'Friend of the House of Numbers'. Constanze told a biographer after her husband's death about 'his love of arithmetic and algebra', as demonstrated by several books of mathematics in his library.

The Magic Flute is not the only opera which illustrates Mozart's obsession with mathematics. *The Marriage of Figaro* starts with a sequence of numbers: 5, 10, 20, 30, 36, 43. These are the opening lines sung by Figaro as he measures out the room that he will share with Susanna once they are married. A curious selection of numbers. But when you add them together you get 144 or 12^2. As with Mozart's letter to his wife Constanze, the square number is a numerical representation of the impending union of Figaro and his bride Susanna.

But is it possible for an audience actually to hear a structure like the golden ratio that Bartók, Debussy and Mozart used as a blueprint for their music? One can see the structure in the score, because it is a piece of geometry; but music when it is played cannot capture this. Music is heard in time, while this geometric proportion is outside time. Mozart used to say that he heard the whole music at once. With such an experience he would have 'heard' this ratio encoded in the overture to *The Magic Flute*. Even without Mozart's extraordinary mind, our brains are still sensitive to seeking out patterns and structure, and hence subconsciously we might well respond to the moment that divides the overture into the golden ratio as the perfect turning point, even though we are not consciously aware of why it feels right.

Silver ratio

Although the golden ratio is king when it comes to aesthetic proportions in the West, there is a different ratio which is favoured in Japan: the silver ratio. This is a more elongated rectangle than the golden ratio and is often the proportion that you find used in Japanese buildings.

A rectangle of proportions A×B, where A is the longer side, is in the silver ratio if the ratio of A/B is the same as (2A+B)/A. This works out as a ratio of 1 to 1+√2. Except for the addition of this extra long side, the silver ratio is defined in a similar way to the golden ratio. If one removes a square of dimensions B×B from this rectangle, then the rectangle that's left over has the same proportions as you find in an A4 piece of paper. The ratio in the A-series paper is 1 to the square root of 2. As I mentioned earlier, this is a clever ratio for paper because when you divide it in two equal parts, for example going from A4 to A5, the ratio is the same.

Just as one finds golden ratios hiding in polygons like the pentagon, the silver ratio crops up in shapes like the octagon. If you compare the ratio of the length of a side of an octagon to the length across the shape, it is in the silver ratio.

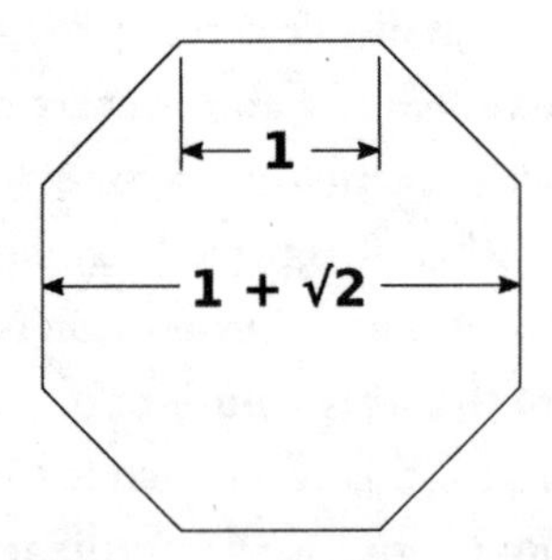

This ratio is sometimes known as 'the Japanese ratio' because it is so favoured by Japanese designers. The design of pagodas often exploits these dimensions. For example, if one compares the width of the top and bottom of the base of the pagoda of the Horyu-ji Temple in Ikaruga, then these two are in the silver ratio.

There has been some suggestion that modern Japanese architects have also tapped into this proportion. The Tokyo Skytree is one of the world's tallest towers. It has two observatories and a digital broadcasting antenna at the top. The total height of the tower is reputed to be in the silver ratio with the distance from the base of the tower to the higher of the two observatories.

One of the beautiful properties of both the silver and the golden ratio is the way one can divide up the shape in some way and get smaller versions of the shape you started with. In the case of the silver ratio, remove a square and you get a rectangle which when cut in half has the same proportions. This is similar to the property that I've highlighted about rectangles in the golden ratio: remove a square and the remaining rectangle again has proportions given by the golden ratio.

By continually cutting more squares from the subsequent golden ratio rectangles, you can build a structure which, when you zoom in, has the same configuration as the original structure. This idea of a shape that retains its identity however much you zoom in on it is actually something that nature exploits a lot, and it is at the heart of our next blueprint.

Blueprint Five
Fractals

In May 2005, the art world thrilled to the exciting news of the discovery of 32 new canvases by Jackson Pollock, one of the leading Abstract Expressionists of the twentieth century. Alex Masters, the son of friends of Pollock, announced that he had uncovered a cache of the artist's paintings when clearing out a family storage locker. It was well known that Pollock had bartered his canvases in transactions with friends in his early years when he was down on his luck. The prospect of unknown paintings by him appearing out of the woodwork was not far-fetched, especially when combined with Masters's parental connection to him. The art market was certainly open to the idea that this could be an incredible addition to the Pollock oeuvre.

Not everyone was convinced. Jackson Pollock had become famous for a very distinctive style, known as 'drip painting'. He would lay canvases on the floor in his studio and then pour paint onto them in great slashes and swirls of colour. It earned him the nickname 'Jack the Dripper'. Although heralded for his unique style by many, he had his critics. There were those who claimed that anyone could toss off such a painting. And this belief has probably fuelled those who have attempted to create fake Pollocks, especially given that a genuine one commands huge sums of money at auction. One of his paintings, *Number 5, 1948*, sold in 2006 for $140 million, making it at the time the most expensive painting ever sold.

So when 32 canvases suddenly appeared on the scene, it raised the question of whether they were really genuine. Was Pollock doing something that was inimitable, or could anyone fake a painting in his style? Pollock himself was very clear. His paintings were not generated by some random toss of the paint pot. 'I can control the flow of paint. There is no accident.' But that still left the question of whether the result had a unique hallmark style that would be hard for a forger to imitate.

Authentication issues had in the past been decided by the Pollock–Krasner Foundation, set up under the will of Pollock's widow, Lee Krasner. The foundation's principal aim is to fund the work of other artists but for a few years it took on the task of cataloguing and validating canvases by Pollock. In 1995, it had disbanded its authentication role. But with this sudden surge of potentially new pieces, it felt the need to play a part in deciding their provenance.

The foundation turned to Richard Taylor for help. The art establishment was somewhat taken aback. Taylor was not a recognised art historian but a physicist. What on earth could a physicist have to offer to the debate? But Taylor had discovered that there was something very distinctive about Pollock's paintings. They were examples of a very important mathematical structure that had only really been identified in the twentieth century. Pollock was painting fractals.

What is a fractal?

Fractals are the geometry of nature. The key to this geometry is that it lacks a sense of scale. What this means is that, as you magnify sections of the shape, the complexity of the shape does not appear to change. If you drew a random squiggle on the page, it might look quite complex on the page; but zoom in on a section of the squiggle at smaller scale and it will simply look like a straight line. A fractal shape, though, has the strange property that if you put several images of the shape captured at different scales side by side, it is impossible to tell which is a close-up and which is the entire image.

The word 'fractal' was coined in 1975 by one of the phenomenon's principal investigators, Benoit B. Mandelbrot. It was meant to have the association of something fractured but also to hint at the idea of fractions. As I shall explain, these shapes have a strange fractional dimension lying somewhere between a one-dimensional line and a two-dimensional plane.

The study of fractals predominately belongs to the late twentieth century, coinciding with the advent of the computer as a tool to explore these shapes. But they were already appearing as interesting mathematical shapes in the nineteenth and early twentieth centuries. Perhaps one of the simplest examples that illustrates many of the characteristics that define fractals was suggested by the Swedish mathematician Helge von Koch.

His principal area of research was the properties of prime numbers but he was intrigued by a geometric challenge that mathematicians were exploring: could you create a curve that at no point had a well-defined tangent. If you take any point on a smooth curve, you can generally draw a unique line called 'the tangent' which touches the curve at that point but doesn't cross it. If the curve you started with is actually a straight line then the tangent at any point is the line itself. Koch wanted to create a curve that was so unsmooth that at no point on the curve could you draw such a tangential line.

To achieve this, he started with an equilateral triangle. The three corner points don't have tangents, because although there are many possible lines through a corner which don't intersect the shape, a tangent has to be a unique such line. But points along the lines do have tangents. So Koch decided to add triangles a third the size of the first one along each of the three edges. This added more points that had no tangents, but still most points were well behaved. But then he wondered what would happen if one kept on adding smaller and smaller triangles along all the straight edges. The shape began to look like a beautiful snowflake and became known as 'the Koch snowflake'.

This is very much a shape of the mind rather than a physically realisable shape. It requires adding ever smaller triangles on the sides. As I shall explore when we come to fractals in nature and art, there are

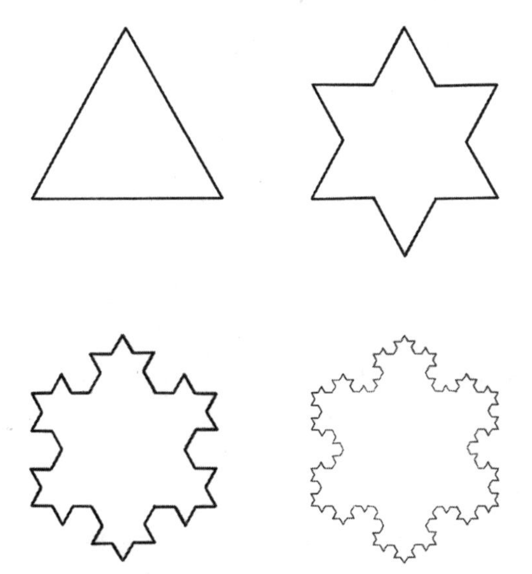

physical limitations which means at some point you would hit the quantum scale. From here, space becomes indivisible and smaller triangles no longer physically exist without the prospect of creating a mini black hole!

But for the mathematician, the idea of infinitely dividing mathematical space rather than physical space is not a problem. Ever since Zeno considered the arrow hitting its target, mathematicians have been exploring the idea of dividing space into infinitely smaller pieces. Koch had no qualms about defining a shape where ever smaller triangles would be added to the sides of the shape.

After adding infinitely many triangles, this shape had the property that Koch was after. The shape was so unsmooth that there didn't exist any point that had a well-defined tangent. But as fractals appeared on the scene in the mid-twentieth century, it was understood that this shape represented a beautiful example of these new geometries.

If I keep zooming in on bits of the shape, then they never simplify. It has infinite complexity. Not only that, the zoomed-in sections look exactly like the shape at a larger scale. It is impossible to tell at what magnification you are looking at the shape. But one of the most curious properties of this shape is the length of its perimeter.

Let's suppose the length of the perimeter of the triangle we started with is one metre. When we added the first set of triangles onto the sides, we increased the length of each side by a factor of 4/3. We divided the line up into three pieces and then replaced the middle piece with two pieces to create the additional triangle. The total length went from three pieces to four pieces, increasing the length by a factor of 4/3.

But every time we iterate this process, taking a line section, dividing into three and replacing the middle piece by another smaller triangle, the length increases again by a factor of 4/3. So at the Nth stage the length is $(4/3)^N$. But hold on. The final shape is made by doing this infinitely often. The conclusion therefore is that the perimeter of the snowflake is infinite in length!

This is very counter-intuitive because the shape fits inside a finite area of space. Fractals can be infinite in length and yet live in a finite amount of space. And this leads to the next interesting property of fractals: they have a fractional dimension. A line is one-dimensional. A square is two-dimensional. A cube is three-dimensional. But these shapes live in the spaces in between.

The Koch snowflake starts life as a one-dimensional line but then tries to spread out to fill the two-dimensional space it is living in. It can't achieve this, but it ends up covering more space than a simple line. Shapes like the Koch snowflake are sometimes called 'space-filling shapes'. Mathematicians have come up with a way to quantify the dimension of these fractals, which has produced a number between 1 and 2. In the case of the Koch snowflake, the dimension turns out to be 1.28.

How to define the fractal dimension of a shape

Take a piece of paper and draw a smooth one-dimensional curve across the page. One way to define the dimension of this curve is as follows. Begin by laying a grid of squares over the top and count the number of squares the curve has crossed. Now increase the mesh of the grid by continually halving the size of the squares. If the curve is smooth, then eventually the number of squares the curve intersects will increase by a factor of 2 each time the mesh gets finer.

If instead we had drawn a closed curve and filled in the centre to make a two-dimensional blob, then eventually the number of squares the shape crossed would increase by a factor of 2^2 each time the squares halved. But something funny happens when we take a fractal, which fills more space than a line but not so much that it is covering the page like a two-dimensional shape. Now the number of squares the fractal crosses increases by a factor of 2^d where d is a number between 1 and 2. If the shape is a line d=1, and if it is a blob then d=2. If it is a fractal, then it is a number in between!

Take the Koch snowflake. Covering the shape with a grid of squares and halving the size of the squares results in the number of squares that the shape intersects going up by a factor of 2^d where d gets closer and closer to 1.28 as the mesh gets finer. It's a shape that sits somewhere between a one-dimensional curve and a two-dimensional blob.

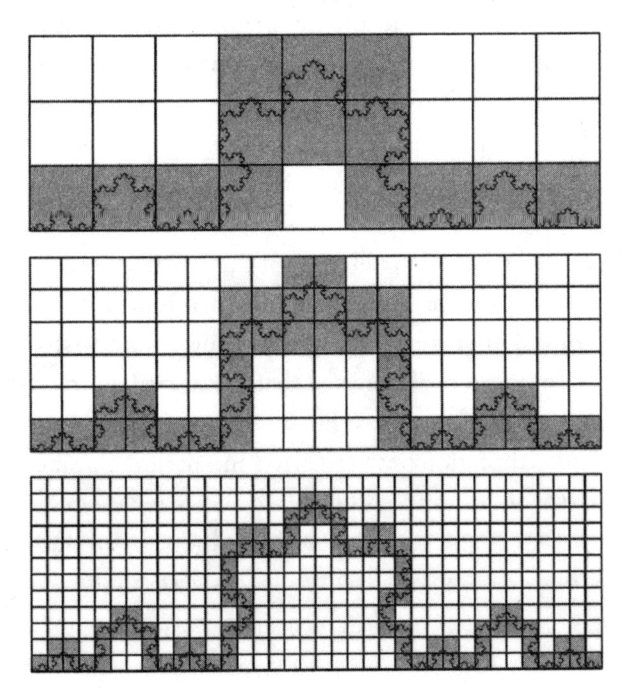

There are different ways to define the fractal dimension of a shape. This fractal dimension is called 'the box-counting dimension' and it is the one that Richard Taylor applied to his analysis of Jackson Pollock's paintings.

Taylor's interest in fractals had stemmed from his work as a physicist. His research focused on the way that electricity flowed through circuits, such as those in a computer. This flow had been well understood in the early years of computer circuitry when although the components were small, of the order of a millionth of a metre, the flow of electricity was still controlled and orderly. But more recently, nano-technology had allowed the size of components to pass a threshold where the flow was no longer so well defined. Instead Taylor was witnessing the electricity taking fractal-like paths through the components, like a lightning strike. If he could understand this fractal behaviour, he believed that he might be able to improve the efficiency of these circuits.

Taylor, though, wasn't just interested in science. He also had a passion for painting, and in 1994 he decided to enrol in a programme at the Manchester School of Art. In the middle of winter, the students were tasked with heading out to the Yorkshire moors to capture the landscape in paint. The weather turned so bad that they had to abandon the project and they retired instead to the comfort and warmth of the college bar to debate the question of how to evoke a snowstorm on canvas.

Taylor recalled an apocryphal story of the artist Yves Klein, who had attached a canvas to the top of his car on a journey from Paris to Toulouse. On the drive, he got caught in a huge thunderstorm. When he arrived in Toulouse, he discovered that the storm had created a rather beautiful pattern on the canvas. He decided to frame it and sold the painting as a collaboration between artist and nature. But it gave Taylor an idea. Get the weather to paint their projects for them. Hang paint pots above a canvas and as the wind blew the pots around the paint would spill onto the canvas. The result was certainly striking. Taylor was staring at what looked like fractal patterns. Not only that. They looked remarkably like Pollock's paintings.

Taylor knew that fractals are very often the geometric signature of chaotic dynamic systems. 'Chaos' is the name given to a system that is very sensitive to small changes in initial conditions. Many processes in nature are chaotic. Water flowing through a pipe, asteroids flying through space, lightning discharging through the atmosphere. If you change the initial condition very slightly, the outcome can be completely different. One of the classic dynamic systems that is chaotic in nature is the weather. It is what gave rise to the expression 'the butterfly effect'. A butterfly fluttering its wings in Brazil might be the tiny change that's needed to turn the beautiful weather in London predicted by the equations into a hurricane.

Weather is chaotic in nature and chaos creates fractals. That is why the paint pots as they got blown around by the storm on the Yorkshire moors were producing images that Taylor recognised from his work on fractals in electrical circuits. But it was this second revelation about the similarity to Pollock's style that began to fascinate Taylor. On his return to the lab, he decided to combine his scientific research with an investigation into whether Pollock was indeed creating patterns that were fractal in nature.

The first characteristic of a fractal is that scaleless quality I've talked about – that if you zoom in on a section, it retains the complexity of the larger-scale image. This is one of the features of Pollock's paintings that perhaps makes them so impressive. I remember one of my first encounters with Pollock's works when I visited the Tate Modern in London. As I stood in front of the enormous canvas and then started to walk towards it, I had this strange sense of getting lost in the painting. It was as if I couldn't tell quite how near I was. The image didn't appear any simpler as I approached it. It was the fractal quality of the painting that was causing this strange effect. Obviously there came a point when I could see the details of the paint that formed the composition. The limits of physics mean that the fractal quality can't continue indefinitely. That encounter with Pollock's painting had a profound effect on me and Pollock has continued to be one of my favourite abstract painters. Little did I realise at the time how much mathematics there was hiding inside the paint.

By itself, though, self-similarity at scale is not enough to make a geometric shape a fractal. After all, the self-similar spirals described in the last blueprint have this property. It also requires that second characteristic I mentioned earlier: the possession of a fractal dimension, the ability to fill more space than a one-dimensional line. To pursue his investigation, Taylor decided to cover Pollock's canvases in the mesh used to define fractal dimension and start counting the squares that the trajectories of paint intersected. As he increased the number of squares in the mesh, he resorted to a computer for help. The hope was to obtain enough data to observe a definite trend in the increase in the number of squares that the trajectories of paint were crossing before the mesh encountered the smallest drops of paint where the fractal disappeared.

The results were striking because the fractal dimension was definitely there. Not only that, the dimension changed across the different canvases that Taylor analysed as Pollock refined his style. Pollock started experimenting with the style between 1943 and 1945, the period that art historians call his 'preliminary phase'. The fractal dimension for canvases created in these years is relatively low, just above the value of 1 that a smooth trajectory would register. For example, an *Untitled* from 1945 scored a fractal dimension of 1.12.

As Pollock experimented more with his dripping technique, Taylor observed the fractal dimension increasing rapidly. Once he hit what art historians regard as his classic period, between 1948 and 1952, the fractal dimension settles at values around 1.7, which is found in paintings such as *Autumn Rhythm: Number 30* (1950). One painting from this period hit a value of 1.89, indicating that the fractals are so dense in character that they are virtually filling the entire space. But Pollock obviously felt that the complexity was too much because he scrubbed the painting, which was done on glass, and took it back to the fractal region of 1.7 that he seemed to favour.

Pollock's approach to creating his paintings began with creating paths of paint on the canvas of approximately 50 to 100 centimetres in length. He would then connect these islands of paint with longer trajectories which started to weave a weblike pattern of paint across the canvas.

This would form the base on which Pollock would then add more and more layers in different colours, often spanning several months of work. What is especially interesting is that Taylor's analysis of the fractal character of the different layers shows that Pollock was refining the fractal dimension of the whole image. It's as if he knew the fractal dimension he was after and he didn't stop adding the paint until he was happy with the effect.

Analysing Pollock's way of painting from footage of him at work certainly indicated that the manner in which the paint was being dripped onto the canvas could well be mathematically chaotic. Often he would pour paint direct from the tin, varying the motion and the height. Other times he would take a brush and flick the paint across the canvas or let it drip. Taylor was interested in whether these techniques by themselves were enough to create a fractal, or whether there was something else that Pollock was doing.

Taylor got 37 students at the University of Oregon to see if they could replicate this fractal quality by splattering paint in a similar style to Pollock. Taylor's analysis of the students' work revealed that not one of them had the fractal quality that his box-counting had identified in Pollock's canvases. It seems that what Pollock was doing was actually quite hard to achieve.

When Taylor applied his analyses to the paintings that had been found in storage that were purporting to be lost works by Pollock it revealed them all to be fake. None of them had the characteristic fractal quality that makes Pollock's paintings so distinctive.

Taylor was intrigued by what was so special about the way that Pollock had painted. He knew that you needed a certain chaotic movement in the way the paint was applied which he'd seen the wind creating in his early experiment on the Yorkshire moors. The mathematics of fluid dynamics indicated that if the viscosity of the paint was within certain parameters, and it was launched in a particular fashion, then the falling fluid could scatter into a sequence of fractal droplets. What Taylor discovered is that these fractal trajectories seem to be achieved by a human body as it tries to recover its balance.

When you watch footage of Pollock painting, it often looks like he is off balance, about to fall. Pollock had serious issues with alcohol, and he wrestled with his drinking problem for much of his life. One of the myths about Pollock is that he painted while drunk, which is what gave rise to the chaotic motion that created the canvases. But the precision and control that his paintings demonstrate give the lie to this myth. He always insisted that there was no accident in his work. Every stroke was intended. His most productive period coincided with two years when he managed to keep sober. Towards the end of his life, when he lost his battle with alcohol, his output dropped dramatically. Alcohol eventually was responsible for his death. He died in 1956, aged 44, in a car accident caused by drunk driving.

Why do people find Pollock's work so appealing? One reason is that the fractals that Pollock painted are very close to the fractals that we find in the natural world. As Pollock explained: 'My concerns are with the rhythms of Nature.' But he never claimed to be painting nature. Rather he simply declared: 'I am Nature.' Taylor's analysis has revealed that Pollock did indeed paint as if he was a force of nature like the wind that buffeted the paint pots that Taylor had erected on the Yorkshire moors.

Some years ago, I had the chance to visit Pollock's studio where he produced all these amazing pieces. The artistic pilgrimage took me to East Hampton on Long Island. As I approached the studio, I was immediately struck by the setting. The studio is surrounded by trees. And trees are a beautiful example of fractal structures. The fractal dimension of much of the natural world is very close to the dimension that Pollock eventually settled on for his most famous and valuable paintings. Pollock had captured in abstract form on canvas the geometry of the natural world that surrounded him. And it is this connection with nature that is probably what makes his paintings so powerful for viewers today.

I am going to let you in on a secret. Inside the studio, I discovered an unlisted Pollock. A Pollock that you won't find in any of the catalogues or galleries of the world. It's staring at you as soon as you enter the space. The floor of the studio is covered in splashes of paint that didn't make it onto the canvases that Pollock had laid on the floor. The floorboards are hidden below layers and layers of paint that Pollock flicked and dripped as he worked away. Once the canvases were removed from the floor, it revealed this incredible unintended painting. I couldn't resist taking a picture. I look at that Pollock every day as it's the picture that I've used as the wallpaper on my laptop. I wonder whether its fractal connection to nature means that it has a calming effect on me as a toil away at my keyboard.

Fractals and nature

It was Mandelbrot's work which revealed that the geometry of the natural world is not the Euclidean, smooth geometry of the ancient Greeks, made up of straight lines and smooth curves, but rather this fractal geometry that emerged in the twentieth century. As Mandelbrot wrote: 'Clouds are not spheres, mountains are not cones, coastlines are not circles, and bark is not smooth, nor does lightning travel in a straight line.' There is an interesting tension here because quantum physics of course implies fractals are a purely theoretical idea that can't exist in reality, just as a circle can't.

At first sight, you might say that the Koch snowflake I created to illustrate fractals still looks very mathematical and certainly too perfect to be part of the natural world. But just adding a little bit of random ness produces something far more natural-looking. At each stage that you add new triangles to the shape, flip a coin to decide whether the triangle sticks out or cuts into the shape. The result is a shape that could easily be the coastline of a country or the contours of a mountain range.

As Mandelbrot discovered, these fractal shapes are everywhere in nature. The florets of a cauliflower. The leaves on a fern. The bronchial network of the human lung. These structures are exploiting a feature of fractals that emerged in our analysis of the Koch snowflake. The infinite length of the outline of the snowflake translates via these fractal structures to shapes that can maximise the space they inhabit. For example, even though the human lung sits inside the relatively small, finite volume of the ribcage, its surface area is huge, and it can therefore absorb a lot of oxygen.

It is interesting that it took so long for these shapes to become part of the mathematician's cabinet of wonders. This is probably due to the fact that computers became a very powerful tool for creating and investigating these structures. The famous Mandelbrot set, which is one of the iconic examples of a fractal, required a computer to reveal its secrets.

My nights out clubbing in the nineties to the sound of Happy Mondays and the Stone Roses were invariably accompanied by images of the Mandelbrot set projected onto walls. The video projections that came with the indie psychedelic music would zoom in on this coral-like structure and yet the shapes would never simplify. I would keep seeing versions of the original shape reappearing inside the fine details of the image, but somehow the video kept plunging deeper and deeper without ever reaching an end. It mirrored perfectly that trippy feeling of falling inside the music that was filling the club.

What is striking is that, without computers, no one would have predicted the complexity that was hidden in the formula that Mandelbrot used to create the image. The computer is to fractals what the telescope is to the exploration of the solar system or the microscope is to the investigation of the molecular world.

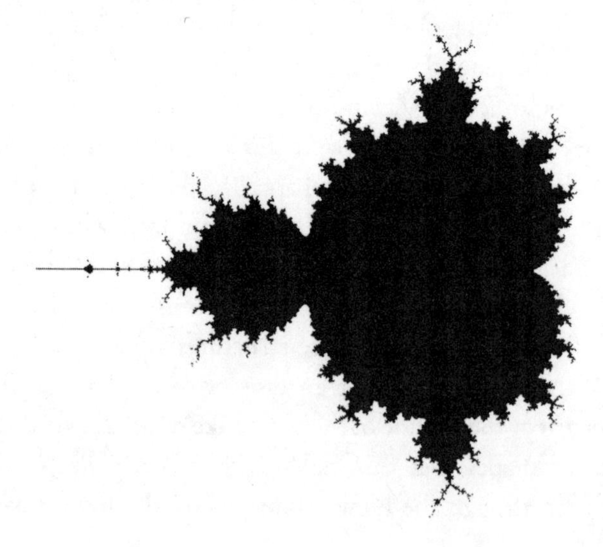

The beauty of the Mandelbrot set is that the key to its creation is an incredibly simple formula. Take a number: x. Square it: x^2. Add a second number, c: x^2+c. So how can this basic recipe give rise to the infinite complexity of this exotic geometry? The secret is to ask what happens if we continually feed the answer we get back into the original formula as a new value for x. Does the answer zoom off to infinity or does it never get far away from where it started?

For example if we set $c=1$ and start at $x=0$, then the iteration gives the sequence: 1, 2, 5, 26, 677, . . . Keep going and the numbers just get bigger and bigger. But set $c=-1$ and start at $x=0$ again, and the iteration gives the sequence: -1, 0, -1, 0, . . . The values just ping-pong between the two numbers, never getting far from the starting point.

The Mandelbrot set is a picture of all the values of c where the iteration does not head off to infinity. So far, we have just taken c to be a number on the number line, with positive numbers running to the right and negative numbers to the left. To get the two-dimensional picture that makes up the Mandelbrot set, we need to allow c to be taken from the larger set of numbers, called 'imaginary numbers', which I first introduced when I explored square roots of negative numbers in my third blueprint. These are built out of the primary imaginary number: the square root of -1, otherwise known as i. This is a number whose square is -1: $i^2=-1$.

What happens if we take $c=i$ in our iterative game? Now the sequence starts: i, $-1+i$, $-i$, $-1+i$, $-i$, . . . Again, it just ping-pongs between these two values. But notice that we have an interesting new number appearing from this formula: $-1+i$. In fact, we can use i to generate a whole slew of new numbers. If A and B are two numbers on the number line, then we can create a new number which is a combination of real and imaginary numbers $A+Bi$. This is called a 'complex number'.

It took until the early nineteenth century for these numbers to be fully accepted as a real part of mathematics. The key to their acceptance was a picture. Mathematicians at the beginning of the nineteenth century understood that you could see these new numbers as points on a two-dimensional map. Regular numbers took you right and left on the number line, but the imaginary number i took you in a perpendicular direction, heading up and down on the map. So the number 3+2i was located three steps to the right and two steps up from the centre of the page.

Now, I can finally define the Mandelbrot set. Take a point A+Bi on this two-dimensional map of complex numbers. Put c=A+Bi in the formula x^2+c and start the iteration at x=0. The point A+Bi is in the Mandelbrot set if the resulting values don't stray infinitely from the starting point but stay within a fixed distance of 0. The incredible surprise is that if you colour in all these points in the Mandelbrot set, then the result is the hugely complex shape depicted on page 164. Zoom in on sections of the shape and you seem to be plunging into hidden worlds, entering an infinite matryoshka doll where each layer is as complex as the last. It's perhaps no wonder that these images became the backdrop to the nineties rave scene.

Even before these fractal projections in clubs and Pollock's drip paintings you can find examples of artists trying to exploit fractal-like structures to capture the natural world. Leonardo da Vinci was one of the first to hint at the role of fractals in the geometry of nature. In his notebooks, he observed the curious way that trees branch and divide, speculating that 'All the branches of a tree at every stage of its height when put together are equal in thickness to the trunk'. The implication being that the tree demonstrates this scaleless property inherent in fractals. The twigs of a tree look remarkably like the large-scale structure of the whole tree. It was the fractals in the trees that Pollock was responding to in his studio hidden in the woods of Long Island.

Leonardo tried to capture the chaotic dynamics of a storm in a series of drawings depicting deluges. To do this, he drew a series of vortices at different scales, smaller ones inside larger ones, hinting at the sense that the shape of the water retains its complexity as one zooms in on the image. Leonardo was obsessed with water, not just from an artistic perspective, but also from a scientific one. His notebooks are full of diagrams trying to describe how one current of water flows into another, or how a current of water behaves when it encounters an obstacle.

Katsushika Hokusai exploited the same idea in his famous woodblock print of *The Great Wave off Kanagawa* created in 1831, which as we saw in the previous blueprint inspired Debussy's piece *La Mer*. The large wave that dominates the picture breaks up into smaller waves which mimic its shape, and these smaller waves in turn are made up of tiny claws of foam that retain the geometry of the overall arch of the large wave. *The Great Wave* is reputed to be one of the most reproduced images in the history of art, reflecting again our human urge towards the artistic representation of the natural world.

If one applies Taylor's strict analysis of fractal characteristics to these earlier examples, then they don't pass the test. They are certainly interesting first forays into shapes that are scaleless, but they don't have the depth and complexity to sustain that fractal quality beyond a couple of iterations. Pollock's paintings exploit the chaos of nature in order to truly represent fractals on their canvases, and they are probably the first paintings to achieve this.

Architectural fractals

The buildings we construct and enjoy living in are often those that mimic the natural world that surrounds us. The success, for example, of Le Corbusier's work is because of its connections to the dimensions of the human body. Given the role of fractals as blueprints for the natural world, the biomimicry we find in buildings means architects have quite often embedded fractal geometry in their designs.

Perhaps the masters of crafting fractals in buildings are the architects of the Islamic world. A building for them is not just inspired by the natural world but also provides a palette to represent the infinite majesty of God. The Muslim artistic tradition very often looks for finite expressions of the idea of infinity. Fractals are a perfect vehicle for this physical representation of the infinite. Even if there are physical limitations on how far one can keep dividing architectural or decorative features, the process implies continuation if physical indivisibility hadn't got in the way.

One of my favourite examples of fractal features in Islamic architecture is the *muqarnas*, a style of ornamental vaulting made up of tiers of repeating cells that vary in size across the different layers. It is often found in the corners of rooms, providing a smooth transition from the walls to the ceiling, which might be domed. Often the interior of the dome will be decorated with this ornate vaulting as well. It resembles some sort of fractal honeycomb and is created by using a fractal-style algorithm. Each larger cell is subdivided using smaller versions of it, which in turn are subdivided into even smaller copies.

I first came across these beautiful fractal features when I visited the Alhambra Palace in Granada while interrailing around Europe as a student. I'd made the trip for the beautiful symmetry of the stucco decoration that covers the palace walls (which I will come to in the blueprint dedicated to symmetry), but I was equally intrigued by these fractal architectural details. They were first developed in eleventh-century Iraq, but they soon found themselves being carved and moulded in buildings across the Muslim world.

One of the finest examples of *muqarnas* can be found in the Taj Mahal. This ivory-white marble mausoleum was built by the Mughal emperor Shah Jahan in memory of his wife Mumtaz Mahal, who died in 1631 at the age of 38 while giving birth to her fourteenth child. When the emperor subsequently died in 1658, he too was buried in the mausoleum, alongside his wife. Inside the central dome of the temple, there is a beautiful example of this complex fractal ornamentation.

When I first visited the Taj Mahal, shortly before heading to a mathematics conference in Assam, I knew in some sense what to expect of the building, having seen so many pictures of it over the years. But it is also part of a much larger complex including a beautiful formal garden. Yet what is often missed from the photos of the perfect, gleaming white mausoleum is that the whole complex is part of the bustling city of Agra. In 2016, Mikhail Yu. Shishin and Khalid J. Aldeen Ismail did a fractal analysis of all these different components of the site: the city plan of Agra, the layout of the garden, the mausoleum itself and finally the ornate fractal decorations that adorn it.

The original garden layout, for example, demonstrates qualities reminiscent of the way the Koch snowflake is constructed. It starts with a square, which is then divided into four smaller squares by two broad paths with a water channel running down their middle. These smaller squares in turn are each divided into four by narrower paths. This is repeated at four different scales, creating an intricate network of paths and green spaces. The design is known as a *charbagh* or *chahar bagh* and is meant to represent the four gardens and four rivers of Paradise mentioned in the Quran.

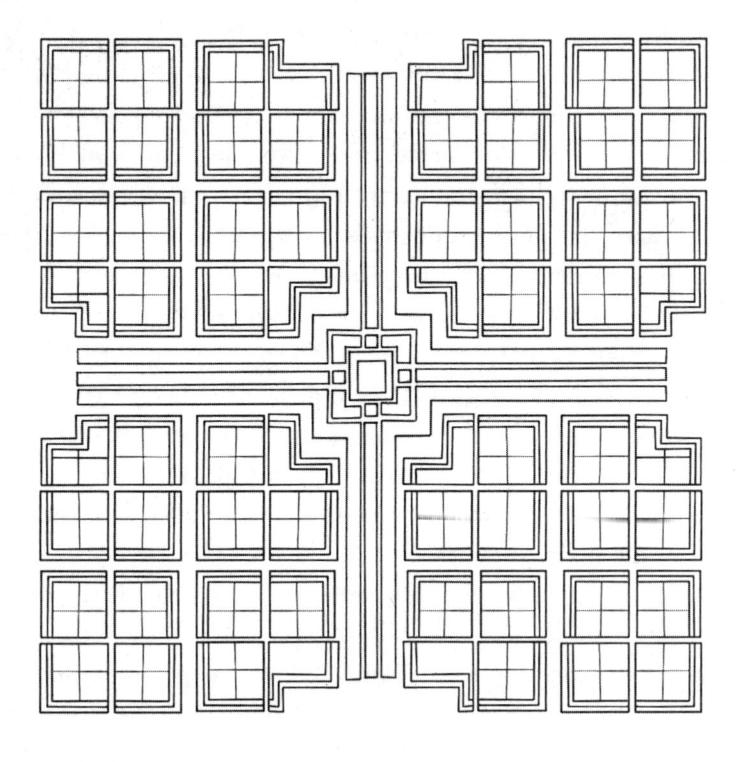

Using a box-counting analysis similar to the one applied to Pollock's paintings, the fractal dimension of the garden's design comes out at quite a high value of 1.8 implying the network of paths is filling a lot of the two-dimensional space of the garden. These fractal characteristics might be linked to the need to distribute water in an efficient way, which is often why one finds fractals in biological systems such as the structure of the lungs or the network of arteries in the body.

The striking result of Shishin and Ismail's research into the Taj Mahal was that there is a close correlation between the fractal dimensions of all the different parts. It's almost as if the whole thing is a fractal that scales all the way from the city layout right down to the fine decorations in the central dome of the mausoleum. This does not appear to be an isolated example. The Iranian cities of Isfahan and Yazd also demonstrate fractal qualities from the small scale of buildings and decorative motifs through to the medium scale of the local neighbourhood and eventually to the macro level of the entire city.

Fractals are also a key component in the design of gardens in Japan. Within the restricted confines that are often home to the traditional Japanese garden, the gardener attempts to create a miniature landscape that evokes the feeling of contemplating the mountains, waterfalls and foliage that make up the topography of Japan.

Japanese gardens often feature lumps of volcanic rock, or *kaseigan*, precisely because they possess a fractal quality which means a close-up photograph of its contours can be mistaken for the outline of a mountain range. The scaleless quality of the geometry of a fractal means that without context it is hard to know exactly how big a piece of volcanic rock truly is. This is why a metre-high rock placed at the heart of a garden can effectively symbolise Mount Horai, the mythical mountain home of the Eight Immortals.

The foliage of a Japanese garden also exploits the mathematics of fractals for its effect. The algorithm for creating treelike structures in computer graphics consists of simply taking a line which then grows, then branches, grows, then branches. Continue doing this, perhaps adding in little random perturbations to ensure the shape isn't too symmetrical and you quickly create something that looks like a tree. It is this fractal shape inherent in trees that is key to explaining some of the choice of plants made by the Japanese gardeners.

One of the appeals of the Japanese art of bonsai is the ability to create a very convincing miniature version of a fully formed tree in a pot that can be kept in a small garden. The fractal shape of trees means that some of these bonsai specimens are extraordinarily convincing versions of their larger cousins, rather like the way a floret of a cauliflower can do a convincing job of looking like a miniature version of the original vegetable. The presence of moss too in a Japanese garden is partly due to the fact that it doesn't give away easily what scale you are at.

This is why water is also so effective in Japanese gardens. The gardener exploits the fact that a tumbling waterfall is controlled by equations of chaos, with the result that the patterns in the water are fractal. This means that once again it is very difficult to tell at what scale you are observing water tipping over the side of a precipice. This ability to replicate in miniature the famous mountain waterfalls of Japan makes cascading water perfect for bringing a Japanese garden to life.

Fantasy fractals

Where fractal images have truly come into their own is in the world of film. Mandelbrot's revelation that these mathematical geometries are the true shapes of nature inspired those who wanted to create artificial natural environments. The use of fractals as a way to simulate nature, combined with the power of computers to implement these ideas, led to a revolution in digitally generated animation and special effects.

One of the first people to exploit fractals in film was computer scientist Loren Carpenter. He was working at Boeing in the late seventies when he read Mandelbrot's seminal book *Fractals: Form, Chance and Dimension*. He decided to test the book's thesis that the shapes of nature were essentially mathematical fractals by creating an animation depicting a mountainous landscape that one could fly through. In the past, such landscapes would be rendered by hand in a laborious process of frame-by-frame painting. Carpenter instead used computers and mathematical algorithms to generate his mountains.

His idea was to repeat the Koch snowflake in three dimensions, using pyramids instead of triangles. Beginning with a very mathematical and abstract landscape of pyramids, he programmed the computer to add on ever smaller pyramids and randomise the process to disrupt the perfect mathematical shapes that would otherwise appear. Even though he'd started the project hopeful of interesting visuals, the quality of the outcome surprised Carpenter. After enough iterations, he had a very convincing range of mountains. Having produced his landscape, he then created a two-minute animation of a flight through the mountains. The result was a film called *Vol Libre*, which has now gone down in the annals of movie history as the beginning of fractal-generated special effects.

Carpenter's film, when it was shown at a computer graphics conference in 1980, prompted a standing ovation and Carpenter was subsequently poached by Lucasfilm, the maker of *Star Wars*. Carpenter's ideas led to the establishment of the Pixar animation studio, where fractals have been used by animators to make the natural backgrounds for jungle scenes like those in *Up* or the Scottish mountains of *Brave*. Half the employees at Pixar are coders rather than artists. This is a world where mathematics and the arts truly work together to create the masterpieces that fill our cinema screens.

With improvements in computing power and mathematical algorithms, fractals are now being used to generate environments for real-life actors rather than cartoon characters to inhabit. Marvel Studios are probably the leading film-makers of these crazy new worlds. My three children's love of the Marvel Cinematic Universe means that it's difficult not to enter a room in our house without bumping into the many different stories Marvel have conjured up. I must admit, I quickly fell under their spell too and ended up bingeing on a large number of the films during the pandemic lockdown. But I was particularly excited to find that my favourites, the *Guardians of the Galaxy* series, also featured some cracking good mathematics.

The team behind the movie *Guardians of the Galaxy Vol. 2* were tasked with creating Ego the Living Planet. Kurt Russell played the planet's corporeal form, but they needed to create the palace and landscape that Ego had built around himself by using his cosmic powers to manipulate matter. The director, James Gunn, and the production designer, Scott Chambliss, both scoured the internet for inspiration for what Ego's world should look like. They were especially taken by the idea of something inspired by the Mandelbrot set, but the trouble was that this geometry was two-dimensional, whereas they needed something that looked like a three-dimensional planet. Was there, they wondered, a three-dimensional version of the Mandelbrot set?

Enter the Mandelbulb. For many years, fractal junkies had fantasised about creating a three-dimensional version of the Mandelbrot set, but there seemed to be an inherent problem. The Mandelbrot set was built by exploiting imaginary numbers, which are naturally represented as a set of two-dimensional numbers, like points on a map. Perhaps if you wanted a three-dimensional version you needed to cook it up with higher dimensional numbers. The trouble was that the mathematics implied you had to skip a dimension and enter four-dimensional space, where numbers called 'quaternions' live. One way to see four-dimensional shapes in our world is to take three-dimensional shadows of these shapes, just as a conventional shadow is a two-dimensional projection of our three-dimensional bodies. But even when one took shadows of the fractals cooked up in this higher-dimensional space, they were disappointing, looking like tangled balls of wool rather than exciting planets that aliens might inhabit.

But mathematicians weren't deterred. Perhaps you could devise a clever formula that took three-dimensional coordinates as inputs and still produce something as exotic as the Mandelbrot set. Rudy Rucker, mathematician and one of the founders of the cyberpunk science fiction movement, made some early attempts in the 1990s. He'd failed to gain tenure in the mathematics department at the university he was teaching at in New York due to his colleagues taking exception to his long hair and the fact that he hung out more with the English and philosophy professors than the mathematicians. His subsequent move into computer science gave him the tools for the first explorations into making a three-dimensional fractal to match the Mandelbrot set.

Rucker came up with a three-dimensional analogue of the formula x^2+c that Mandelbrot had used, but the results were disappointing, never quite producing the exotic landscapes to match the Mandelbrot set. The computer graphics in the 1990s were also not really up to the job. His experiments did give rise to a rather trippy short story that Rucker wrote called 'As Above, So Below: A 3D Mandelbrot Story'.

Just as the early computer was an essential tool for Mandelbrot to uncover the two-dimensional fractal that bears his name, navigating the realms of three-dimensional fractals had to wait for the arrival of technology that could truly get mathematical explorers to the limits of these worlds. That took another 20 years. But it also required a new mathematical idea. In 2007, software designer Daniel White had been playing around with a squaring formula and new computer graphics tools to experiment with what might be hiding inside these three-dimensional fractal shapes. But it was engineer Paul Nylander who wondered if it was the squaring that was holding back the exotic potential of this formula in three dimensions. He started experimenting with taking higher powers of input number: x^3, x^4, x^5. The magic started to happen when he took 8th powers.

The Mandelbulb of order 8, as it is known, was a revelation. It was indeed like landing on some exotic alien planet. As White and Nylander used the graphics tools that were now available to render these formulas, texture and complexity akin to the Mandelbrot set in two dimensions began to emerge. They discovered a world full of complex cave structures, exotic gardens populated with intricate flower-like shapes and ornate Baroque edifices that looked like fractal cathedrals. Their work inspired a whole new generation of artists exploring these striking fractals.

It was some of these images that the Marvel Studio special effects team had stumbled on in the mid-2010s and realised that they had found their Ego. Some of the fractal artists using these formulas to generate worlds were brought onto the project. Kevin Smith was one of the team asked to make these fractal worlds a virtual reality: 'Visual effects for me has always been about the combination of art and science and it was great to be able to take a purely mathematical concept like fractals and not only make something new and different, but to use it to help drive the narrative of an awesome movie like *Guardians of the Galaxy Vol. 2*.' The special effects created with fractals helped earn the film its nomination for 'Best Visual Effects' at the 2018 Oscars, but it lost out to *Blade Runner 2049*.

While the Mandelbulb was perfect for creating the alien landscape of Ego, it was another quality of fractals that the team making *Doctor Strange* wanted to exploit. They loved the way that as you dive deeper and deeper into a fractal you seem to be entering other possible worlds that never simplify. The Mandelsponge, a variant of the bulb, proved perfect for what they wanted to do.

Kaecilius, the sorcerer pitched against Doctor Strange, is meant to be able to twist and manipulate reality. His mastery of the Dark Dimension allows him to fracture space-time into shards that he can fight with. Fractals were perfect for this splintering. He is also able to manipulate and unfold space. The city of New York, home to Doctor Strange, ruptures to reveal smaller copies of itself hiding inside. Buildings bifurcate, walls ripple, streets split and spiral to reveal inner worlds. Fractals provided the feeling there are universes living inside universes that was ideal for the narrative of *Doctor Strange*. At one point, even Doctor Strange's hand turns into a fractal hand, with smaller hands branching off his fingers, and even smaller hands branching off the fingers of those hands.

It is the striking balance that fractals capture between the reality of nature and the imaginary world of mathematics which means they are a perfect medium for creating the Marvel Cinematic Universe that my kids (and I) so love.

Listening to fractals

Fractals are essentially visual artefacts. They are the visual representation of chaos. The geometry of nature. But is it possible to hear a fractal? What would a musical representation of a fractal sound like? While visual fractals play around with space, a musical fractal seeks to exploit time as its medium for scaling. In a visual fractal, we see the same geometries appearing at different scales. Zoom into a Jackson Pollock canvas and you encounter the same intricate patterns in the details of the painting as you see from afar. Fractal music plays the same game but at different timescales. What you might expect to hear is the same musical motif repeated at different speeds, elongated and contracted.

Amazingly, this musical idea of a fractal is already present in some of the very early Renaissance polyphony of composers such as Josquin des Prez and Johannes Ockeghem. 'Polyphony' just means music with two or more independent voices singing together; a classic form of it is the canon, which we encountered when exploring the circles in Bach's Goldberg Variations. In a conventional canon one voice starts singing and then is joined by a second voice singing the same melody but delayed in time. However, Josquin des Prez and Ockeghem also developed something called a 'prolation canon'.

In this type of canon, voices often start together but they sing the same melody at different speeds. For example, in the Agnus Dei of Josquin des Prez's *Missa 'L'homme armé' super voces musicales* written in the late fifteenth century, the soprano sings the same melody as the alto but three times as fast. The soprano singer can repeat the melody three times before the alto has finished their rendition of the tune. Underneath this you have a third voice, the tenor, which sings the melody at double speed, completing two cycles before the slowest voice finishes. The same motif is sung at three different speeds, in a 3 to 1 to 2 ratio.

On the page, this even looks rather fractal-like. The same pattern can be seen but at different scales. The intriguing thing about music is that one cannot experience the whole piece in its entirety in one instant in the same way that you can look at a painting and see it all. You have to wait for the notes to be sounded as the musicians sing through the composition. It's as if you're looking at something through a vertical slit which allows you to see only a small sliver at any one time and only gradually scans right across it. It would be very difficult to detect fractal imagery in this manner.

That's why those interesting geometric features used to create music are often easy to detect on the page but hard to hear in the playing. They often require a musical memory of what has just passed in order to be able to appreciate them. And this is probably why they are used, because they engage the brain in recognising and anticipating patterns as it listens to the music. The brain is not just present in the here and now, but is examining the past and predicting the future.

Another fan of these fractal canons is Bach. No surprise there. The 14th of the puzzle canons that he wrote to accompany the Goldberg Variations consists of a single voice, but then the algorithm says to expand this into a piece for four voices where each one sings at twice the speed of the one beneath it, so that we hear the same motif sung at tempos that are in an 8 to 4 to 2 to 1 ratio. It's as if you zoom in at four different temporal magnifications and at each level the same pattern can be heard. Just as with a Jackson Pollock painting, there are limitations to the true fractal nature of these representations. A fractal expects you to zoom in infinitely small and still see the same complexity. Both paint and notes have limits to their divisibility, so four levels of scaling is probably the limit of what might work for a musical fractal.

The Goldberg Variations themselves demonstrate a certain fractal quality. There are 32 movements in total, consisting of two renditions of the aria and 30 variations in between. But each movement is made up of 32 bars of music. Each zoomed-in movement is like a microcosm of the whole piece.

The idea of these fractal compositions had a renaissance in the twentieth century with a number of composers creating pieces that explored musical motifs at different speeds. Perhaps the most elaborate example occurs in Shostakovich's Fifteenth Symphony. In the first movement, three phrases played by the string instruments occur at speeds in a complex ratio of 8 to 6 to 5. The result is a very intricate counterpoint that is notoriously challenging to perform.

There is another intriguing way that composers have experimented with embedding fractals into music. Instead of multiple voices layered to create a motif running at different tempos, this is a fractal created in a single voice. Rather like the Koch snowflake, one starts with a motif or musical phrase inside which we are going to create the fractal. Each individual note of this musical phrase is substituted with the entire same musical phrase transposed to start on the note it is replacing and that is then played quickly enough to finish before the next note of the original phrase. We hear the same pattern of notes, but shifting up or down in pitch according to the notes of the original melody.

Just as with the Koch snowflake, one could repeat this again and replace each note of the faster motifs with even faster copies. The effect is to hear the same motif at different speeds, as in a prolation canon, but somehow nested inside each other to make up one line of music. In reality, a composer is unlikely to go beyond just one iteration of this process. A common way of realising this idea is to take a sequence of, say, four notes from an ascending scale and replace each of them with a run of four notes starting on the corresponding original note of the scale and played at four times the original speed.

Literary fractals

If musical fractals are already stretching the imagination, what about the concept of a literary fractal? Tom Stoppard's *Arcadia* is often held up as the Maths Play. The dialogue certainly has lots of discussion of mathematical ideas, like Fermat's Last Theorem, chaos theory and, indeed, fractals. Stoppard admits it was reading James Gleick's popular science book on chaos and fractals that became the inspiration for a play now regarded as a modern classic.

I went to see *Arcadia* when it was first staged at the National Theatre in the 1990s and was very impressed by how, in addition to the witty exchanges about the mathematics, Stoppard actually allowed the audience to experience the play's ideas by embedding the mathematics into the way the drama unfolds.

Arcadia is a chaotic experiment enacted for the audience. As one of the characters in the play explains, chaos theory asserts that although the future might be determined by mathematical equations, these can be very sensitive to small changes in the initial conditions. One tiny alteration can cause an enormous difference in the outcome. The setting of the play in two time periods – the early nineteenth century and the late twentieth century – lets us see the effect that a small perturbation in the drama during the earlier period has on the future.

The young Thomasina, the central character of the nineteenth-century drama, has made an amazing mathematical discovery. As she says to her tutor Septimus: 'You will be famous for being my tutor when Lord Byron is dead and forgotten.' But we know from our modern perspective that history took a different direction. It turns out (spoiler alert) that Thomasina dies in a fire on the eve of her seventeenth birthday, meaning that her mathematical discoveries are lost. Yet we see that a small change in the drama could so easily have saved her. If her tutor Septimus had actually accepted her sexual advances and joined her in her bedroom that night, he might have saved Thomasina from the fatal fire. Small changes have big impacts. Byron is famous. Septimus is forgotten. The butterfly effect enacted on stage.

The mathematical discovery that Thomasina makes relates to the geometry of fractals. Alas, due to her early death we had to wait until the twentieth century before mathematicians rediscovered this new sort of shape. But the wonderful thing is that Stoppard has succeeded in making his play an example of a fractal

In *Arcadia*, the modern strand of the drama spans a few days and represents a telescopic version of the nineteenth-century story, which extends over several years. Time and again, we see patterns in Thomasina's story repeated in the twentieth-century narrative. They are not exact copies but perturbations like the zoomed-in sections of a fractal. Thomasina's mathematical brilliance is mirrored by the modern mathematician Valentine, Byron by the literary critic Bernard; and, in a beautiful fractal climax, we see two different couples in two different time periods waltzing together through the same room, creating what one might describe as a Stoppard fractal to match Mandelbrot's.

For Stoppard, one of the attractions of writing a play about fractals was the chance to explore the fact that mathematics, nature and the creative energy of the arts were all running in parallel. 'There is a complementarity in the notion that math describes nature, and that nature is following mathematical rules . . . Numbers have a kind of social behaviour, they're not simply tools of description,' Stoppard says.

The other play I've seen which embodied its theme in its structure so successfully is Michael Frayn's *Copenhagen*. First performed in 1998, it stages the meeting that took place in 1941 between the quantum physicists Niels Bohr and Werner Heisenberg. No one knows quite why Heisenberg made the visit to Copenhagen. Was he warning the Allies via Bohr about the Nazis' plans to develop a nuclear weapon? Or was he seeking help to solve the equations that would lead to the successful development of such a weapon? In the play, Frayn revisits their meeting over and over, and each time the interaction is different. Frayn used Heisenberg's uncertainty principle as a metaphor for how an episode in history can change every time you look at it.

The many-worlds interpretation of quantum physics posits that at every moment in time the universe splits into parallel universes, each of which follows a different path. In his 2012 play *Constellations*, Nick Payne took this idea and beautifully wove together a multilayered narrative in which an encounter between two people is rerun multiple times until finally the couple find a universe in which they are together. Both *Copenhagen* and *Constellations* are like theatrical experiments in quantum physics. In his earlier play *Hapgood*, Stoppard too explored the narrative potential of quantum physics as a great metaphor for the world of espionage, with quantum particles replaced by agents and double agents.

Fractals, some believe, are the very blueprints of narrative. The television producer John Yorke, in his seminal book about storytelling in drama, *Into the Woods*, identifies the fractal as an important structure. A drama often has a three-act shape. The beginning, the middle and the end. The set-up, the crisis, the resolution. *Cinderella*: girl meets boy, girls loses boy, girl re-finds boy. But stories are built from acts that are built from scenes that are built from even smaller units known as 'beats'.

Each of these components has its own three-act structure. For Yorke, a piece of drama very much has the shape of a Koch snowflake. Triangles inside triangles inside triangles.

This idea of fractal triangles inspired part of the structure underpinning David Foster Wallace's 1996 cult classic *Infinite Jest*. The novel gets its name from Hamlet's words: 'Alas, poor Yorick! I knew him, Horatio: a fellow of infinite jest, of most excellent fancy!' A sprawling multilayered narrative, it was acclaimed by *Time* magazine in 2005 as one of the 100 best English-language novels published since 1923.

Novels don't usually have footnotes. But *Infinite Jest* is a novel where even the footnotes have footnotes, and this is where we start to see a layered, fractal narrative emerging. The fractal structure seems to be working in the main narrative too, where ideas are introduced briefly, then reappear as longer passages that also contain other brief ideas, which in turn recur at ever greater length. When an interviewer picked up on this fractal structure, Foster Wallace was somewhat taken aback. He'd never admitted to this blueprint before:

> I've heard you were an acute reader. That's one of the things structurally going on, it's actually structured like something called a Sierpiński gasket, which is a very primitive kind of pyramidal fractal. Actually, though, what was structured as a Sierpiński gasket was the draft that I delivered to [my editor] in '94, and it went through some, I think, mercy cuts, so it's probably kind of a lopsided Sierpiński gasket now.

A Sierpiński gasket shares things in common with the Koch Snowflake. It starts with a solid triangle, but then an inverted triangle one third the size is removed from the centre. The same action is then performed on the three triangles remaining inside the shape, and again on the smaller triangles just created, and so on. The repeated removal of shrinking triangles produces this fractal shape, called 'the Sierpiński gasket'.

The Koch snowflake started with a one-dimensional outline, and by adding infinitely many shrinking triangles we created a shape with a dimension larger than 1 but smaller than 2. The Sierpiński gasket goes the other way round. By infinitely removing triangles from the solid triangle, the shape drops from a two-dimensional solid shape to a fractal with dimension 1.585. It is named after the Polish mathematician Wacław Sierpiński, who described the structure in 1915. The 'gasket' part of the name was given by Mandelbrot, who thought the shape resembled the gaskets you find inside car engines, which consist of a series of decreasing holes designed to stop oil leaking.

But the Sierpiński gasket's shape had actually been around long before these mathematicians started exploring its fractal properties. You find these patterns of decreasing shapes in latticework and in mosaics created by artists in the twelfth and thirteen centuries, especially in Italy. It is known as the 'Cosmati style' after a family of artists operating in Rome who used this fractal as a blueprint in many of their mosaics.

That a novelist like Wallace should know about this mathematical shape may seem surprising, but he was always an author who was equally fascinated by mathematics and literature. He studied literature and philosophy at Amherst College in Massachusetts and specialised in mathematical logic as part of his course. It was during this period that he encountered Georg Cantor's ideas of infinity, which became something of an obsession. He even wrote a non-fiction account of them in 2003, called *Everything and More: A Compact History of Infinity*. Mathematics would be an ever-present blueprint in the way that he would structure narratives.

If Jackson Pollock's canvases are made from pixels of paint that come together to create fractal structures, could there also be authors using words as if they were the pixels of a novel to create texts that have a fractal quality? For example, an author might vary the use of long and short sentences across their novel to create a fractal shape. This would have an important effect on how the novel reads. The way the author uses sentence length in a passage will affect the fluency and rhythm of the text. Interspersing long sentences with short sharp bursts has a very different feel to a paragraph where all the sentences are of the same length. This local pattern might then be reflected in the larger scale structure of the novel, for example in the length of paragraphs or chapters.

In 2016, a team of Polish physicists based in Cracow took 113 novels written in English, French, German, Italian, Polish, Russian and Spanish, from Virginia Woolf's *The Waves* to Tolstoy's *War and Peace*, in order to analyse how the length of sentences varied as one read through them. By looking at the text, at different scales they could analyse how the long and short sentences were distributed.

They discovered that the distribution of sentence lengths in *War and Peace* had a reasonably low level of complexity. When the analysis was done on *The Waves*, however, it was found that the distribution of short sentences was much more complex than those of the longer sentences. The opposite turned out to be the case for Julio Cortázar's novel *Hopscotch*, a book I will return to when I consider the blueprint of randomness: it was the distribution of the longer sentences that was more complex. But perhaps unsurprisingly, the novel that demonstrated fractal complexity across all sentence lengths was James Joyce's *Finnegans Wake*. Joyce once said that he wrote it 'to keep the critics busy for 300 years'. The analysis revealed that if a novel's style resembled a stream of consciousness, like *Finnegans Wake*, then the fractal complexity of its sentence lengths would be high.

For Joyce, the structure of his work was extremely important.

> I am really one of the greatest engineers, if not the greatest, in the world besides being a musicmaker, philosophist and heaps of other things. All the engines I know are wrong. Simplicity. I am making an engine with only one wheel. No spokes of course. The wheel is a perfect square. You see what I'm driving at, don't you? I am awfully solemn about it, mind you, so you must not think it is a silly story about the mouse and the grapes. No, it's a wheel, I tell the world. *And* it's all *square*.

It may not be obvious from this description quite what is going on underneath the bonnet of Joyce's work, but it's clear that something mathematical if not entirely logical seems to be inspiring his prose.

From the twentieth-century geometry of fractals, we return to geometries that are some of the most ancient and ubiquitous in the mathematical and natural world. These symmetrical solids also have had a fascinating role to play in shaping our creative output.

But before we leave the world of fractals, I hope you are now sufficiently up to speed for me to share one of my favourite mathematical jokes. What does the 'B' in Benoit B. Mandelbrot stand for? Benoit B. Mandelbrot. You should never have to explain a joke, but just in case you are left cold, it makes his name into a fractal, because however much you zoom into what that B stands for you've always got another B lying one layer below.

Blueprint Six
The Platonic Solids

Rudolf von Laban had been rehearsing his dancers for months. It was a huge honour and responsibility to prepare a large-scale piece for the opening of the Berlin Olympics in 1936. The authorities in Germany were proud to showcase the work of a man who'd become the leading choreographer in Europe.

Laban had revolutionised dance in the decades leading up to this important commission and was universally recognised as the driving force for a new theory of movement that was sweeping through the world of choreography. He'd witnessed the massive impact that painters such as Picasso and Kandinsky had had on visual art and was keen to channel that revolution into movement theatre. His new style of choreography moved away from dance telling a story, playing second fiddle to the music, and instead revelled in the body as an abstract structure moving in space. It was Laban's style of dance that had informed the piece that I had created with the choreographer Carol Brown to represent the proof of the irrationality of the square root of 2, which appropriately enough we had performed at the Trinity Laban Conservatoire in London.

Laban's early studies in architecture played an important role in his understanding of structure, and it was geometry in motion that became the blueprint for his unique style of dance. His other revolutionary idea was the belief that contemporary dance was not the preserve of elite dancers but was for everyone. His idea of community choreography led to the invitation to create a grand spectacle to launch the Olympic Games.

The piece, entitled *Of the Spring Wind and the New Joy*, would be performed simultaneously across 30 cities by 1,000 young dancers on 31 July, the day before the Berlin Olympics were due to start. It was going to be a wonderful celebration of the power of dance to 'take people out of their everyday experiences and – perhaps unconsciously – feel themselves merging with the cosmos'. Laban had been greatly influenced by Nietzsche's writing and part of the text of *Thus Spake Zarathustra* would be read alongside the dance. As Nietzsche's Zarathustra said: 'And we should consider every day lost on which we have not danced at least once.'

One of the principal ideas behind Laban's innovative new theory of dance was the awareness of space and structure surrounding a dancer. For Laban, the idea of geometric shapes surrounding the body were key to guiding the movement of the dancer. Laban was inspired in part by reimagining Leonardo's *Vitruvian Man*, an image that he'd encountered in his architectural studies. Instead of being enclosed in a two-dimensional square or circle, a dancer has a third direction to explore: back and forth.

Laban had first experimented with getting dancers to imagine the lattice of a cube surrounding them, but he found that the cube was strangely unsuited to the natural movement of the body. The icosahedron, with its 20 triangular faces meeting at 12 points, turned out to be much more tuned to the human form. As he wrote, 'Man is inclined to follow the connecting lines of the twelve corner points of an icosahedron with his movements travelling as it were along an invisible network of paths.' The far simpler tetrahedron of four triangular faces meeting at four points joined by six lines also proved a natural shape for dancers to explore.

These geometric shapes gave rise to an important new form of notation that Laban had developed which constituted instructions for where each of the limbs of the body should move. The corners of the geometric shapes were labelled with different symbols and the notation could be read like a musical score. But instead of the time line extending horizontally, as with a musical score, it ran vertically up the page. The length of the symbols represented time spent in a location, and each vertical line corresponded to a part of the body, like the different instruments of an orchestra.

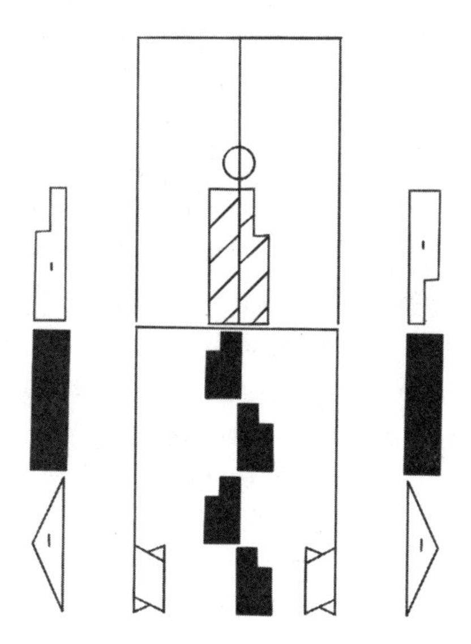

Each symbol corresponds to a different position, as illustrated here for the right leg.

It was with the aid of this new notation that Laban was able to coordinate dancers across 30 cities.

Laban's ideas were hugely successful and led during the period after the First World War to 24 Laban schools being set up across Europe to promote his new geometric style of dance. But as the Weimar years turned into the years of National Socialism in the 1930s, politics started to impact on the running of the schools. There is evidence that Laban started to implement the policy of removing non-Aryan students well ahead of official requirements to exclude students that didn't match up to the Nazis' Aryan credentials. A letter by Laban written in 1933 records that he 'has removed all non-Aryan pupils from the children's course effective from the end of the season. End of season: 7 July 1934. Attached please find the new pupil register.' It wasn't until 1938 that such a requirement became law.

By 1934, Laban was being entrusted by Joseph Goebbels with organising major dance festivals to promote Nazi propaganda. He was appointed director of the newly established German Dance Theatre set up by Goebbels's ministry. Laban wrote during this time of how 'we want to dedicate our means of expression and the articulation of our power to the service of the great tasks of our People. With unswerving clarity our Führer points the way.' Laban was walking a precarious tightrope, trying to keep Goebbels onside while still trying to promote a more universal message through his dance.

The piece Laban devised for the Olympic opening ceremony outwardly had all the trappings of Nazi propaganda. Dancers were dressed in identical neoclassical costumes adorned with the symbol of the sun wheel rather than the Nazi swastika, or *Hakenkreuz*. The *Sonnenrad*, or sun wheel, was one of a number of medieval European symbols appropriated by the Nazis in their attempt to invent an idealised 'Aryan' heritage.

Of the Spring Wind and the New Joy began with a depiction of the horrors of war, before a passage reflecting on the emergence of a new Germany. This gave way to increasingly joyful celebrations culminating in the dancers joining with the audience in a circle of prayer and unity depicting the *Sonnenrad* writ large. This was all accompanied by Laban's reworking of the words of Nietzsche's *Zarathustra*:

> Is it only in dance that I find a way of talking about higher things? Since there have been people, man has not rejoiced enough. This above all is our inheritance from the earth, my brothers! If we learn to have more pleasure, then we will unlearn best how to do others harm and to imagine how to strive.

The dancers came together on 20 June for a full dress rehearsal in front of a high-ranking audience including Goebbels, who'd commissioned the piece. The reception was not good. Goebbels hated it. Outwardly it looked like Nazi propaganda, but its message was far from clear. What was really being celebrated here? Germany was moving to a war footing, where fighters were to be lauded not seen as a mistake of the past. The sacrificial death of the country's young was part of Nazi ideology. But Laban, who had lived through the horrors of the First World War, was trying to evoke a time beyond war.

The piece was far too intellectual, allowing viewers too much freedom to interpret it for themselves. It just wasn't unequivocally a celebration of National Socialism with the Führer at its heart. Following the dress rehearsal, Goebbels confided his unfavourable impressions to his diary: 'Dance rehearsal: freely based on Nietzsche, a bad, contrived and affected piece. I forbid much of it. It is all so intellectual. I do not like that it goes around dressed in our clothes and has nothing whatever to do with us.'

Goebbels's banning of the performance marked the beginning of Laban's gradual downfall in Germany. The Nazi regime started investigating his political and personal background. When attempts to cast doubt on his Aryan ancestry failed, the Gestapo switched their attention to his former membership of the Freemasons. Laban had indeed been the Grand Master of a Masonic lodge in 1917–18, but had left the order before the Nazis came to power and banned the organisation. But the Gestapo kept hounding him. Why had he not joined the National Socialist Party? Had he been abusing state funds in his rather chaotic management of the German Dance Theatre? There were even questions raised about his sexuality.

Laban desperately tried to hold on to his position in the face of all this negative attention, but eventually he got the message. He offered his resignation on 31 March 1937. His attempts to find further employment in Germany were blocked by Goebbels. Reluctantly Laban decided to flee to Paris, but eventually it was England that would become the home for his ideas about the role that mathematical shapes like the icosahedron can play in choreography.

The icosahedron, however, wasn't a blueprint solely for Laban's movement theories. It has a very ancient heritage as possibly the blueprint for the whole of creation.

Nature's building blocks

Laban's icosahedron is one of the shapes that the ancient Greeks believed were the building blocks of the physical universe. It was thought that matter was built from four basic elements: earth, wind, fire and water. In his dialogue *Timaeus*, Plato suggested that each of these elements corresponded to a symmetrical shape.

The shape of earth was the stable cube. Air corresponded with a shape comprising eight triangular faces meeting at six points, called an 'octahedron'. The shape of fire was the tetrahedron, a triangular-based pyramid whose four faces are equilateral triangles. It is the spikiest of all the symmetrical forms, so is well suited to being the shape of the element of fire. In contrast, the smoothest of the shapes is the icosahedron, with its 20 triangular faces. This is closest to the spherical shape of a drop of water and explains why Plato chose it to represent the fourth element.

The first mathematical exploration of the shapes appeared in Euclid's *Elements*, where he proved that the only symmetrical three-dimensional solids were these four, plus one more. This fifth shape was made up of 12 pentagonal faces meeting at 20 points, and it is called a 'dodecahedron'.

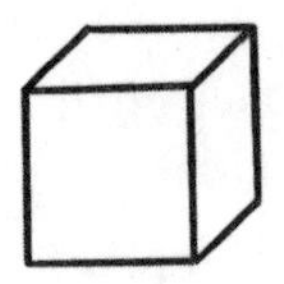
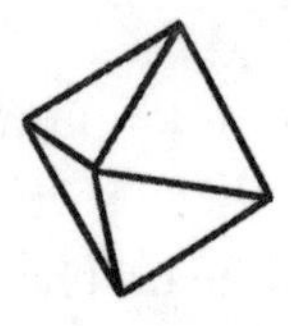
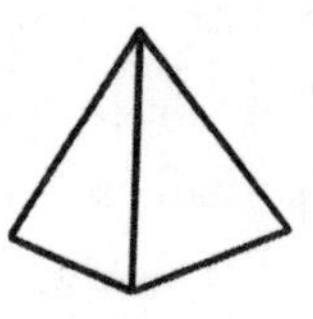
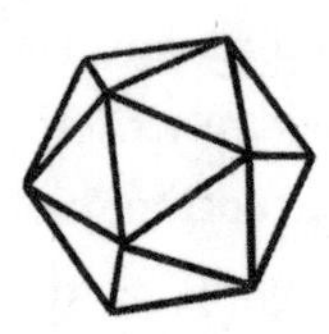
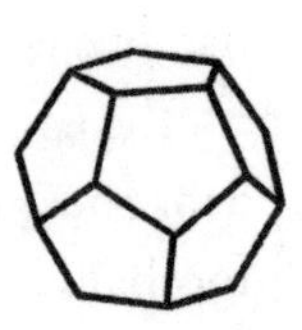

Euclid proves why it is impossible to make a sixth shape out of identical symmetrical faces put together in a symmetrical fashion, in the sense that the number of faces that meet at a point is the same for each point of the shape, ensuring all the vertices look identical. The proof is the highlight of the *Elements*. The rest of the text seems to be gearing up to this mathematical finale. The five shapes have become known as 'the Platonic solids', after Plato's discussion of them in his dialogue. In addition to the four shapes that had been assigned to the four elements, Plato associated the remaining shape, the dodecahedron, with the structure that enclosed the universe.

Even if ancient Greek chemistry based on earth, fire, air and water might not have made it into modern science, the recognition that these shapes are some of the most basic blueprints for the natural world has survived. Cubes and tetrahedrons are key to the construction of crystals, including salt and ice. Viruses from herpes to Covid are biological icosahedrons. The skeletal structures of certain protozoa take the form of octahedrons, as beautifully illustrated by nineteenth-century biologist Ernst Haeckel in his book *Kunstformen der Nature*, known in English as *Art Forms of Nature*. The soil that covers the earth includes silicates that are in the shape of a tetrahedron, which might have been a better choice by Plato for the shape of earth rather than the cube.

Although the mathematical analysis of these symmetrical shapes had to wait until the ancient Greeks, there is a lot of evidence that early artists were already exploring their potential. My local museum in Oxford, the Ashmolean, is home to a fascinating collection of carved stone balls that were discovered in Scotland (the British Museum and the National Museum of Scotland also have impressive holdings). They date back 5,000 years to Neolithic times. Each is about the size of a fist, but what makes them special are the carvings on them.

The balls are made up of circular patches arranged in as symmetrical a manner as possible – much like the leather pieces with which modern footballs are fashioned. The most common designs consist of four or six circular faces, which correspond to the symmetry of the tetrahedron and the cube. But some of the sculptors were intrigued to see if there were other arrangements that worked. There are balls with eight circular patches that match the eight faces of an octahedron, but this was definitely a period of experimentation rather than mathematical rigour. You can find stones attempting to arrange seven or nine patches. The Ashmolean has a ball with 14 of them. In the National Museum of Scotland, there is a ball with 53 small knobs, and balls have been found with as many as 160 divisions.

What was the purpose of creating these objects? If you want to make dice for playing games, then you'll need shapes with symmetry. You find several of the Platonic solids being used to play some of the very earliest games that humans developed. The dice found with the so-called 'Royal Game of Ur' – a Sumerian board game dating to roughly the same period as the Neolithic stone balls – are tetrahedrons. Cube-shaped dice with pips corresponding to scores of one to six have been excavated from sites belonging to the ancient Harappan civilisation of the Indus valley, today part of modern Pakistan. But the Neolithic balls found in Scotland do not appear to be dice and are not believed to be part of any game. Indeed, their function is still something of a mystery.

Some have suggested they were symbols of status for tribal leaders. Others have proposed that they were used for divination. Many of the balls have beautifully intricate spiralling lines carved into their faces similar to designs found in Neolithic tombs in Scotland and Ireland, such as the Newgrange tumulus in County Meath. Two groups of three intertwining spirals arranged in a triangle greet you as you enter the tomb. Known as a *triskele*, one group is a mirror image of the other.

Perhaps we are making a mistake in looking for utility in these stone balls. Instead, maybe they are an expression of the artistic and mathematical character that was already burning strongly in the people of Neolithic times: exploring symmetry for its own sake, wanting to discover new possibilities or seeing the limitations of three-dimensional shapes. It is interesting that the artist came first, while we had to wait 2,000 years for the mathematicians to play their part in proving that there are really only five ways to arrange these patches in a truly symmetrical manner.

Renaissance symmetry

We see the Platonic solids making another artistic appearance during the Renaissance. Three-dimensional symmetrical shapes start popping up in numerous artistic works of the period. On the table in the foreground of Jacopo de' Barbari's portrait of Luca Pacioli, the mathematician who wrote the treatise on the golden ratio, we can see a perfectly rendered dodecahedron. Suspended on Pacioli's right is a strange glass shape made up of a combination of triangular and square faces. On the floor of Saint Mark's Basilica in Venice there is a mosaic of a stellated dodecahedron often attributed to the artist Paolo Uccello. This is a star shape where each face of the dodecahedron has been replaced by a five-pointed star. Uccello also depicted a woman in a fresco of the biblical Flood sporting a strange geometric hat in the shape of a polyhedral torus.

The same shape appears in the intarsias, or wooden mosaics, that Fra Giovanni da Verona enjoyed making. In a series of panels he created for the church of Santa Maria in Organo in Verona, there is a whole array of symmetrical shapes, including icosahedrons, stellated dodecahedrons and a spherical lattice made up of hexagons and pentagons in the configuration many will recognise from the classic shape of a football. In Dürer's famous engraving *Melencolia I*, as well as the 4×4 magic square, we find a sphere and a large strange shape made from triangles and elongated pentagons. The Flemish painter Nicolas Neufchatel depicted the mathematician Johann Neudörfer and his son exploring the properties of a dodecahedron while a cube lattice floats above them.

Why this sudden explosion of interest in all these different geometric shapes? Basically the artists of the Renaissance were showing off. This was the period when artists had cracked the theory of linear perspective. Suddenly they understood how to faithfully represent the three-dimensional universe on a two-dimensional surface. One way to show off their skills in perspective was to paint a perfectly symmetrical three-dimensional object. The symmetry meant that the shape was very unforgiving. Any small error in representing the symmetry and the illusion of perspective would be broken. More organic shapes like trees or faces would tolerate a certain amount of inaccuracy. But not these mathematical figures.

Luca Pacioli, the individual depicted in de' Barbari's portrait, helped to reinvigorate the study of symmetrical shapes. He wrote a number of influential books that described the current state of mathematics at the end of the fifteenth and beginning of the sixteenth centuries, including his treatise on the golden ratio. Although principally a mathematician, he was also sensitive to the world of art, believing that his subject was indispensable to it. A quote often attributed to Pacioli declares: 'Without mathematics there is no art.'

He was certainly renowned for his expertise in the use of mathematics for creating a true sense of perspective, and artists would come to learn from him how to implement his theories. In his 1509 book *De Divina Proportione*, he explored the ways that many of these symmetrical shapes encoded the golden ratio in their proportions. This was the book that was full of illustrations drawn by the friend that Pacioli made while teaching mathematics at the ducal court of Milan: Leonardo da Vinci.

You really couldn't ask for a more masterful illustrator of these mathematical shapes than the great Leonardo. His diagrams are beautiful. They include not only images of the solids but also beautifully rendered frames where the faces are removed and just the edges remain, so that you can see through the shape. They probably depict models Pacioli constructed himself out of wooden struts. In Leonardo's illustrations, they hang from strings like Christmas baubles.

In addition to all the Platonic solids drawn by Leonardo, Pacioli's book also features some interesting new symmetrical shapes. For example, there is an illustration of the shape that forms the basis of the classic football, made up of 12 pentagons and 20 hexagons that Fra Giovanni da Verona had also depicted in his wooden mosaics. The edges all have the same length, and they are put together in such a way that all the vertices are identical, with two hexagons and one pentagon meeting at each point. This is called a 'truncated icosahedron' because you can create it by cutting off the corners of an icosahedron. That's why it's a good football, because it's an attempt to make the icosahedron even more spherical.

For a mathematician, this shape immediately prompts the question: how many more symmetrical shapes are there if you relax the condition that all faces must be identical? The answer was discovered by Archimedes, who found that it was possible to make 13 different shapes from combinations of two or three types of symmetrical face put together in a symmetrical fashion.

No text by Archimedes survives that documents his discovery of what are now called the 'Archimedean solids'. We only know that he deserves the credit thanks to a book written some 500 years later by another Greek mathematician. In his *Mathematical Collections*, Pappus of Alexandria mentioned that Archimedes had discovered 13 ways to build these multifaceted shapes, and he described how they were put together. But visualising them is almost impossible from what Pappus said. Given that the mathematical theory of linear perspective was still more than a thousand years away from being developed, it's perhaps not surprising that Pappus' text didn't come with helpful illustrations.

For me this is what is so striking. We actually had to wait for artists who were skilled enough to represent the three-dimensional universe on a two-dimensional surface for these shapes finally to emerge from the depths of time. Often my story is a tale of artists being inspired by the objects that fill the mathematician's cabinet of wonders. This is a beautiful example of artists helping mathematicians to see for the first time shapes that had once been conjured up in the mind of Archimedes.

The artists who started exploring these new shapes in the fifteenth century were actually doing so without knowledge of Pappus' book, which appeared in print for the first time only in 1588 (although some fragments had been available since about 1560). One of the first was Piero della Francesca, who is better known for masterpieces such as the *Flagellation of Christ*, a stunningly complex work finished around 1460, which displays his absolute command of the new theory of perspective. The painting is famous for the way it foregrounds three large figures in one half of the scene while locating Christ in the background of the same space in the other half. Its perfect rendering of the illusion of depth is based on the mathematical understanding of how the sizes of the different figures relate to each other.

Piero was able to master the challenge of creating perspective accurately because his other obsession was mathematics. He was not alone in this: the biographies of many artists of the time mention their skills as mathematicians. One has to remember that the education system during this period had evolved out of the medieval 'quadrivium', which taught the four disciplines of arithmetic, geometry, music and astronomy. The concept of a liberal education has its origins in an era when students were expected to master the arts and the sciences in tandem.

Piero was the author of several mathematics books, including the *Trattato d'abaco*, or *Treatise on the Abacus*, which actually includes a huge amount of geometry. The book is full of challenges to the reader to calculate how long certain edges are on various shapes that he describes. For instance: 'There is a spherical shape whose diameter is 6. I want to put in it a body with eight faces, four triangular and four hexagons. I ask what its edge is.'

However, it is the shapes themselves that are intriguing. This particular problem is accompanied with a drawing which shows the shape is composed of triangles and hexagons. The sketch he made isn't exactly a great example of Piero's skill at rendering three-dimensional shapes in two dimensions, but it is nonetheless clearly recognisable as one of the 13 Archimedean solids.

Whether Piero had read about this shape somewhere or discovered it for himself is unclear. But he realised that one way to make it – and other new symmetrical shapes – was to chop off the corners of the Platonic solids. In this case, he obtained it by removing the corners of a tetrahedron. Hence its name: the 'truncated tetrahedron'.

Piero's books of geometry were the beginning of the reawakening of the ancient shapes that Plato, Euclid and Archimedes had talked about. Piero described six of the 13 Archimedean solids, five of them obtained by this process of amputating corners. It appears that Pacioli could well have read the *Trattato d'abaco* and then, without crediting it, reproduced many of the shapes that appeared in its mathematical challenges. But he came up with a couple of new forms that weren't in Piero's books. With Leonardo as illustrator, it was perhaps inevitable that Pacioli's text became the go-to source for those interested in these emerging new shapes.

One of Pacioli's discoveries appears in the guise of the strange glass shape that hangs in de' Barbari's portrait of him. It is made from 18 squares and 8 triangles, and it is called a 'rhombicuboctahedron'.

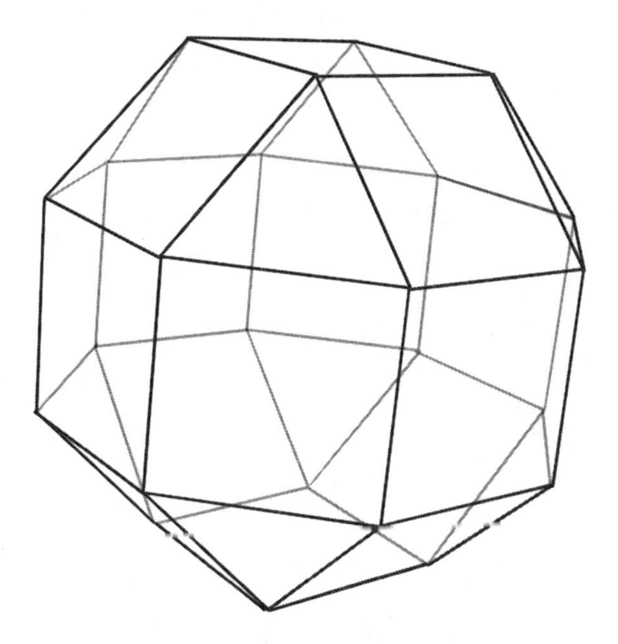

There is some evidence that Leonardo might have helped de' Barbari render it so perfectly, since he'd already successfully drawn one in Pacioli's book. Some have even suggested that Leonardo himself had discovered it, because Pacioli incorrectly stated that it could be built by removing the corners from other symmetrical shapes. One can speculate that Leonardo came up with the idea while doodling and then Pacioli appropriated it by trying to show that it could be realised in a similar fashion to all the other Archimedean solids.

In his *Divina Proportione*, Pacioli talked about the symmetrical shapes he presented being just a selection from what he believed was an infinite number of possibilities. Addressing his patron Duke Ludovico Sforza of Milan, to whom he dedicated the book, he stated:

> It does not seem to me, most noble Duke, that I should extend my discussion of these bodies, aware as I am that their progression goes on indefinitely by the continued and successive cutting off, one after the other, of their solid angles, and according to this their differing shapes come to multiply.

What is apparent from this is the fact that mathematicians at the time were still not aware of Archimedes' discovery, and Pappus' account was not generally available until much later.

Piero's and Pacioli's strategy of making new symmetrical shapes by cutting the corners off the five Platonic solids and seeing what new forms might emerge yielded only six of the 13 shapes discovered by Archimedes, although they did also succeed in working out another two. But it was not until Albrecht Dürer became involved that any more were identified.

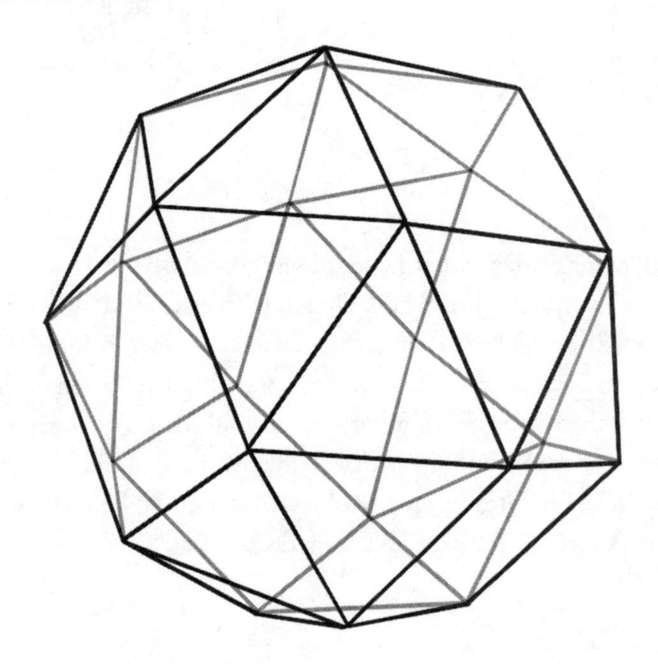

In 1505, Dürer travelled to Italy from his home in Germany in order to learn about the new art of perspective. At some point during his stay, he bumped into Pacioli; indeed, he could well have been seeking out Pacioli as a renowned teacher of the mathematical theory behind the subject. It seems their conversations were not limited to learning the skills of perspective but also explored some of the new geometric solids that were emerging. Their discussions on truncated shapes led Dürer to discover two more possibilities: the truncated cuboctahedron and the snub cube. He eventually documented them in 1525, in his *Underweysung der Messung mit dem zirckel un richt scheyt*, or *Instruction in Measurement with Compass and Ruler.*

Dürer's treatise was the culmination of his mathematical investigations and was aimed at addressing those German artists who, he believed, 'have grown up in ignorance, like unpruned trees'. But as the artist declared in his introduction the work was 'not only for painters, but also for goldsmiths, sculptors, stonemasons, carpenters, and all those for whom using measurement is useful.' Rather than illustrating the shapes in it with his well-honed skills of perspective, Dürer instead represented them in the form of two-dimensional nets. For example, the snub cube is shown as a flat plan (known mathematically as a net) made from six square faces and 32 triangular ones. It's possible that Dürer discovered it by using these nets as a device to experiment with making new shapes, because this shape can't be produced by simple truncation.

What is particularly exciting about all this is seeing how some of the leading artists of the Renaissance – Piero della Francesca, Leonardo da Vinci, Albrecht Dürer – employed their artistic skills in order to participate in this journey of mathematical discovery. As we saw with the Neolithic artists, experimentation is very good at realising what might be possible. But to prove that you have found them all requires the mind of a mathematician.

It was the German astronomer and mathematician Johannes Kepler who developed the new approach that, in 1619, finally yielded the description of all 13 shapes as well as the proof that no more were possible. In

Proposition 28 of Book 2 of Kepler's *Harmonices Mundi*, or *Harmony of the World*, we find the first recorded list of all the Archimedean solids. The names we use today for these shapes derive from those that Kepler gave them in his description. His breakthrough was to go back to the idea that Euclid used to prove there were only five Platonic solids. This involved placing faces around the vertices of potentially new symmetrical shapes and seeing what resulted. Leonardo probably built the rhombicub-octahedron in this way. Actually, Kepler perhaps could have benefited from a collaboration like the one Pacioli had with Leonardo. The illustrations for the Archimedean solids in the *Harmonices Mundi* were all made by the astronomer Wilhelm Schickard, but they reveal that he was probably a better scientist than he was an artist.

It is possible that Kepler was beaten to the discovery of all 13 shapes. In 2006, a set of 40 woodcut printing blocks came to light in the storage depot of the Albertina Museum in Vienna. Dating to the mid-sixteenth century, the blocks depict nets for many symmetrical solids. Of these, 10 are explicitly Archimedean solids, while the missing three are hinted at by extra marks that are contained in the other woodcuts. Who was the mysterious block-cutter who deserves the credit for beating Kepler to the rediscovery of the shapes that Archimedes had cooked up all those years ago?

Three of the blocks feature the signature of Hieronymus Andreae, who was an artist and engraver who worked alongside Dürer. But most of them were probably cut by a craftsman in Andreae's workshop. The state of the blocks reveals that only a few prints were ever taken from them. Perhaps the plan was to use them in a new book about the shapes, but that seems never to have materialised.

Dalí, the carnivorous fish

Leonardo's *Last Supper* is one of the great examples of Renaissance painters wrestling with the challenges of perspective. The point at infinity, defined by the lattice of squares that make up the ceiling, is located precisely at Christ's head, focusing the viewer's attention on the most important figure in the work. But more recently, the *Last Supper* has also provided the setting for a painting by one of the great artists of the twentieth century which explores the intersection between art and the Platonic solids.

Salvador Dalí is a painter who constantly found inspiration from the world of science, and especially mathematics. He famously declared: 'I am a carnivorous fish, swimming in two waters. The cold water of art and the hot water of science.' He much preferred to invite scientists rather than artists to the soirées that he hosted. When asked by a journalist from *Le Figaro*, 'Why such a great interest in science?', Dali replied that artists really didn't interest him. They often failed to have any idea of scientific progress and for him this was the key to discovering new ways to view the world.

Like many students, I enjoyed covering my digs with Dalí's paintings of melting clock faces, thinking I was being terribly deep by displaying these artistic depictions of Einstein's theory of relativity. But despite his love of mathematics, I'm not really a great fan of Dalí. Although I love the celebration in his paintings of all things scientific and mathematical, they often feel a little bit forced, as if he is trying too hard to show off his scientific credentials. Nonetheless, his Surrealist style of painting melding the worlds of art and science, combined with his talent for performance, made him one of the most popular artists of the twentieth century.

Dalí's exposure to the world of science began while he was a student in Madrid. He lived at the Residencia de Estudiantes, which hosted events by leading thinkers of the time, including Le Corbusier, Einstein, and Stravinsky. Dalí didn't finish his studies, as he was expelled in 1926 for insulting one of his professors during his final examination. But the impact of listening to the likes of Einstein and Le Corbusier had a big impact on the themes that he wanted to explore in his art. A trip to Paris to work with Picasso some months after being expelled inspired his exploration of the ideas of Cubism and its attempts to visualise space beyond the limits of our three dimensions.

His initial interest was the inner psychological world and the work of Freud, which inspired many of his Surrealist paintings. Freud was his guiding light. The images that Dalí conjured up in the paintings he made during this period were meant to capture the hidden unconscious drives that Freud talked about in his work that were responsible for human behaviour.

But as he learnt more science he was drawn to the exciting emerging ideas of fundamental physics. He turned from the interior world of the personal to the exterior world of cosmology and the quantum. The ideas that were emerging during this period seemed to transcend the psychological narrative. This was a language to explain how the whole universe worked. Freud was left behind. As he declared: 'My father today is Dr Heisenberg.'

Despite the attraction of new ways of seeing and understanding the universe, Dalí still believed there was a lot to learn from the artists, scientists and especially the mathematicians of the Renaissance. As he wrote in his 1948 book *50 Secrets of Magic Craftsmanship*: 'I tell you here, young painter, YES, YES, YES and YES! you must, especially during your adolescence, make use of the geometric science of guiding lines of symmetry to compose your pictures.' Despite the attraction of all the new structures that were emerging in modern science, he was adamant that this needed to be underpinned by a complete mastery of the geometry of the Renaissance.

But wouldn't such rigid mathematical scaffolding be the death of artistic creativity? The romantic artist didn't want to spend his or her time worrying about geometric details that just got in the way of true artistic expression. On the contrary, responded Dalí, 'it is in order not to have to think and reflect upon them that you make use of the properties'. During these years, he dedicated all his efforts to reading and mastering the mathematics contained in Pacioli's *De Divina Proportione*. So perhaps it isn't surprising to find Dalí, seven years later, not only painting the *The Sacrament of the Last Supper* but including an intriguing mathematical framework for his rendition. Christ and the Twelve Apostles are seated inside a huge transparent dodecahedron. Its faces are windows through which you can see the rugged cliffs and boulders that Dalí's home in Catalonia looked onto. It is a view that crops up in several of his paintings. An ethereal, disembodied torso with outstretched arms, which represents the Creator, floats above the scene, enmeshed with the struts of the dodecahedron.

Plato had singled out the dodecahedron as the shape of the cosmos. But, as Dalí explained, there was a lot of other numerological significance going on inside the picture beyond just the twelve pentagons that made up the faces of the dodecahedron. Christ was accompanied by twelve apostles. Twelve was central to the way we mark time: twelve hours of the day, twelve months of the year, twelve constellations in the night sky. As he wrote, 'I wanted to materialise the maximum of luminous and Pythagorean instantaneousness based on the celestial communion of the number twelve.'

The painting is laid out according to proportions that reflect this Pythagorean influence. The physical dimensions of the canvas were deliberately chosen to be in the golden ratio, a concept that Dalí might have heard Le Corbusier discussing when he visited the Residencia de Estudiantes in Madrid. The arrangement of the figures inside the painting also reflects this ratio. For example, the two disciples either side of Christ divide the width of the painting in the golden ratio. The height of the table for the supper when compared to the height of the painting is set at the golden ratio. The dodecahedron would also have appealed to Dalí because its proportions conceal many golden ratios, as Pacioli discussed in his treatise.

The Sacrament of the Last Supper coincided not only with Dalí's longstanding fascination with science but also his turn towards Catholicism and religion, which was ignited by the horrors of the Second World War. He described himself as 'a real madman, living and organised with a Pythagorean precision', and his work at this time increasingly embodied a strange amalgamation of the spiritual and the scientific – a style that he neatly epitomised as 'nuclear mysticism'.

Tesseracts

Platonic solids had already featured in another of Dalí's most popular paintings combining mathematics and his religious interests. But instead of the three-dimensional shapes that Plato had studied, Dalí used one that exists in four dimensions. *Crucifixion (Corpus Hypercubus)*, completed in 1954, depicts Christ crucified on a four-dimensional cube.

Given that it was already a challenge to depict three dimensions on a two-dimensional canvas, you might ask how Dalí was able to represent shapes in four dimensions. What is depicted is actually a four-dimensional cube unwrapped into three dimensions. Just as Dürer had represented his new shapes by creating two-dimensional nets that could be printed on the page of his book, it is possible to create three-dimensional shapes that represent the nets for creating four-dimensional forms. Of course the problem is that we are limited to our three-dimensional universe, so we can never see the actual shapes built with these nets. But the nets do at least give us a glimpse of what might exist out there in a four-dimensional universe.

The net of a three-dimensional cube consists of six squares joined together in a cross shape. Perhaps this image of a cross is what gave Dalí the idea for his painting. The net of the four-dimensional cube in *Crucifixion (Corpus Hypercubus)* consists of eight cubes, four stacked vertically and four arranged around the faces of the third cube in the stack. It looks like two intersecting crosses. For Dalí, the sense of a dimension beyond the physical resonated with the spiritual world that he was trying to capture in the painting. He referred to this painting as an example of 'metaphysical, transcendent cubism'.

This unwrapped hypercube is sometimes known as a 'tesseract'. It is a shape that has inspired not just artistic representation but also literary interpretation. Author Alex Garland's second novel is named after the shape, but for Garland the tesseract is intended to act as a symbol for a missing God, not a spiritual dimension as Dalí had used it.

Garland came to fame for his first novel, *The Beach*, published in

1996. It told the story of a group of backpackers who discover a secret beach in Thailand, but gradually this paradise turns into a frightening dystopia. I loved this book. Many reviewers commented on how the book was a wonderful mix of *Lord of the Flies* and John Fowles's novel *The Magus*, another book I'd loved as a student. It was regarded as giving a voice to the emerging Generation X, a generation that I can just claim to be part of. It quickly became a cult classic, made into a movie starring Leonardo DiCaprio, and it shot Garland to international fame at the age of 26.

Second novels after such an astounding debut are tricky beasts. So I was very intrigued to see a mathematical shape chosen for the title. *The Tesseract* continued Garland's obsession with South-East Asia. Set in Manila, it told the stories of three characters whose lives seemed completely independent until chance caused them to violently collide. For Garland, the city of Manila always felt like a place with many different polarities and independent dimensions. The narrative is much less linear than *The Beach* and is structured using the idea of this unwrapped hypercube. Given the highly fractured nature of the narrative he needed a structure that could keep the story coherent. It was the shape of the tesseract that provided the key.

The unwrapped hypercube gives one an oblique way of exploring this four-dimensional shape that always remains out of view. The three spatial dimensions that we inhabit allow us a partial sense of its overall shape. Garland has taken this idea but replaced dimensions with narratives. There is an overarching narrative to the novel, but none of the three characters whose stories are told can see it all, only a part of the whole. By the end of the novel, though, the reader has navigated these narrative shadows and gained a sense of the hidden shape that they reveal.

While the missing fourth dimension for Dalí was meant to hint at a spiritual realm, for Garland the opposite was the case. 'It's a book about atheism,' he says. Garland's idea was that reality didn't need a God to explain it. You just needed mathematics. This is a novelist after my own heart. Garland says he gets frustrated by people who fill the unknown and unseeable with God.

In a radio interview he talked about the tendency of those who feel that there must be some overarching principle at work in the universe to endow it with agency and consciousness, to call it a god. But for Garland the opposite was the case. The existence of mathematical blueprints as an explanation for the underlying structures we experience replaced any need for some mystical conscious being to be conjured up to explain them. The fact that the mathematics might transcend physical reality didn't endow these structures with consciousness.

Writing a novel according to the blueprint of a tesseract was not easy. He admitted that he was not prepared for how complicated keeping track of a narrative that had this multi-dimensional structure was going to be. He was keen to be true to its mathematical character but found that sometimes this meant that he lost sight of the actual story he wanted to tell in order to be faithful to the mathematics. 'It was a very strange and compulsive exercise,' he said.

Garland is at heart a storyteller who loves narrative and character, so he succeeded in his task of marrying this with his passion for structure. But he remembers his editor's response when he sent them a book whose title was a strange mathematical shape from the fourth dimension. 'The blood drained from the publisher's face as I handed it over.'

Personally, I really enjoyed the novel and thought it very cleverly exploited this four-dimensional blueprint to structure the story. But then I am a math nerd. Its wider reception in the end was not great. Many were disappointed with this rather over-intellectualised second novel. Garland returned the advance for a third novel and moved very successfully into writing scripts for movies, where he continues to explore the edges of science with films such as *Ex Machina*.

Dalí too continued to be obsessed with the idea of representing the mathematical universe beyond our physical three-dimensional world. In 1979, he painted a canvas explicitly called *Searching for the Fourth Dimension*. It was a passion that he'd witnessed Picasso wrestling with when he'd visited him in Paris decades earlier. Cubism attempted to represent a multidimensional world by layering many two-dimensional perspectives within a single canvas. Picasso's explorations in representing multidimensional space had been inspired by reading about the work of the nineteenth-century German mathematician Bernhard Riemann, who had been one of the first to find a pathway from the physical three-dimensional universe into the abstract world of hyperspace.

How to see in four dimensions

Bernhard Riemann is best known for his work on the distribution of prime numbers, but his other passion was for geometry. The secret to his passage to worlds beyond our three-dimensional universe was the discovery of a numerical code that formed a blueprint for physical space. Riemann understood there was a dictionary that translated geometry into numbers, and vice versa. This was one of the great breakthroughs made by René Descartes in the early seventeenth century. Geometry before Descartes very often entailed drawing pictures to tease out the properties of lines and circles. Descartes was keen to create a language that could allow you to explore the world of geometry without being taken in by any strange illusions that it might hide. He once said that 'sense perceptions are sense deceptions'. He believed that, by using numbers, it was much harder for mathematicians to be deceived.

This desire to change slippery geometry into transparent numbers led to the development of what is now called 'Cartesian geometry', after its inventor. On a two-dimensional page, any point can be identified by two numbers that encode how far left or right, up or down, it is. Often we draw a set of axes with measurements along the lines, which can be used to help identify the coordinates of every point.

If you draw a square, then you can translate this picture into numbers by giving the coordinates for the square's four corners: for example (0,0), (1,0), (0,1) and (1,1). Two-dimensional space is translated into coordinates with two numbers indicating how far to move in two independent directions. We can extend this dictionary translating geometry into numbers to describe shapes in three dimensions by adding a third coordinate, which encodes the height above the page that your shape might sit. To encode a cube in this way requires eight coordinates, each with three numbers. The cube sitting above the square that we drew has vertices at the corners given by the following coordinates: (0,0,0), (1,0,0), (0,1,0), (1,1,0), (0,0,1), (1,0,1), (0,1,1) and (1,1,1). One can think of making the cube by taking the square in two dimensions described by the first four points, and to create a second square hovering about it simply sliding the first square into the new third dimension by lifting it one unit off the page, as encoded by the 1 in the third entry of the last four points. As I explained in the blueprint dedicated to the circle, this sliding of shapes from one dimension into another is how Borges describes the process of making shapes in higher dimensions in 'The Book of Sand'.

Descartes stopped at three dimensions, content that his dictionary captured the geometric world in numbers. It was Riemann, two centuries later, who realised the potential for one half of this dictionary to be extended in order to describe geometries beyond the three dimensions we can see. Although the geometric side of the dictionary reached its limit at three dimensions, Riemann saw no reason not to just keep adding coordinates to the numerical side. Each new coordinate would correspond to a new hypothetical direction independent of the three we can see around us.

In this way, Riemann ushered in the era of hyper-dimensional geometry. The numerical side of the dictionary provided a language to articulate a hypercube in four dimensions. You simply add a fourth coordinate which encodes moving the three-dimensional cube in this new direction, giving 16 coordinates in total – those for the original cube: (0,0,0,0), (1,0,0,0), (0,1,0,0), (1,1,0,0), (0,0,1,0), (1,0,1,0), (0,1,1,0), (1,1,1,0); plus eight new coordinates corresponding to the three-dimensional cube hovering in hyperspace above it: (0,0,0,1), (1,0,0,1), (0,1,0,1), (1,1,0,1), (0,0,1,1), (1,0,1,1), (0,1,1,1), (1,1,1,1).

With this language, one can investigate the properties of this hypercube and establish that it is made up of 32 one-dimensional edges, 24 two-dimensional squares, and 8 three-dimensional cube-shaped faces that are pieced together in four dimensions to build the hypercube. One can actually see these faces in Dalí's *Crucifixion (Corpus Hypercubus)*, because these cubes survive when we unwrap the shape into the three-dimensional net which the painting depicts. Dalí paints the eight cube-shaped faces stacked four high with four more making up the crosspieces.

These hyper-dimensional worlds aren't just the playthings of mathematicians or the inspiration for artistic creativity. It turns out that they are the blueprints for the very fabric of the universe. Once again, evidence of the three-way conversation between mathematics, art and nature at work. When physicists started exploring the implications of the fact that light travels at the same speed regardless of your frame of reference, a Euclidean three-dimensional view of the universe started throwing up strange paradoxes. It took Einstein's insights to sort out the mess. His breakthrough was to frame the universe within a four-dimensional geometry of space-time that gave him the arena in which to formulate his theory of relativity. Einstein was always struck by how the geometry he needed to describe the way the universe worked had been conjured up half a century earlier by mathematicians interested in the beauty of geometry for its own sake without any need for utility to motivate their research.

We've already seen a hint of this four-dimensional world in the blueprint dedicated to the circle. The equation for a circle $x^2+y^2=1$ maps the coordinates (x,y) of points which are a fixed distance from the centre. The equation for the sphere $x^2+y^2+z^2=1$ does the same for points in three-dimensional space. Riemann's language then simply extended this into the world with four independent directions in order to produce the equation for a hypersphere $x^2+y^2+z^2+w^2=1$.

Borges had been fascinated by the books he'd read that tried to explain such mathematical transformations from shapes you can see to shapes of the mind. He had used narrative as his tool for exploring these shapes, while Dalí chose to represent them visually by unwrapping them into the three-dimensional world. But it is only with the mathematics of Riemann that we can truly 'see' these shapes of the mind.

Catastrophe theory

Dalí's obsession with mathematical ideas as blueprints for his paintings would continue right up to his death. He was a fervent advocate of the importance of artists and scientists spending time together, so he was very excited by the prospect of a symposium organised by the University of Barcelona in 1985, entitled 'Culture and Science: Determinism and Freedom'. It was held under the dome of the Dalí Theatre-Museum in his home town of Figueres and explored the role of chance in science and art, a theme that we will return to in our final blueprint. Dalí was keen to attend but his ill health by this time meant that he was restricted to watching the debates from his bedroom via a video link. Eager to join in the conversation, he invited some of the attendees to his room, including the mathematician René Thom.

Dalí had met Thom some years earlier and had become enthralled by his research into catastrophe theory, a geometric approach to the mathematics of chaos. Thom was interested in using mathematics to describe natural phenomena such as the shifting of tectonic plates or the sudden movement of a landslide where small changes in the system can lead to catastrophic results. It also helped with understanding phase changes in the process of morphogenesis, which causes cellular tissue in an organism to develop its shape.

Thom's research sought to classify the different types of catastrophe that might occur. This led to identifying seven scenarios which describe dynamic systems of two variables determined by four parameters. These were surfaces wrapped up in various ways in four-dimensional space which could describe the behaviour of different natural phenomena. The names that Thom gave these catastrophes probably helped pique Dalí's interest. They included the 'swallowtail catastrophe' and the 'butterfly catastrophe'.

Despite the fact that Thom was awarded a Fields Medal for his work, he was always regarded as something of a maverick by the mathematical community. In their first meeting together, the two eccentrics recognised kindred spirits and forged a lasting friendship. When Dalí understood that these catastrophes could describe natural phenomena such as the sudden movement of tectonic plates, he shared with Thom a crazy theory he had about plates meeting at the railway station in Perpignan. He believed that this was the location of the centre of the universe. According to Dalí, Thom didn't dismiss the theory as the idea of a madman, but responded with his own ideas about the fault-line in the Pyrenees that caused the fold mountain range near Perpignan.

Dalí's interaction with the mathematics of catastrophe was responsible for several of his last paintings. In *Topological Abduction of Europe – Homage to René Thom*, the fracture through the earth at the railway station in Perpignan is portrayed as it splits apart. In the lower left-hand corner is the equation $V=x^5/5+ux^3/3+vx^2/2+wx$, which describes the geometry of the surface in accordance with the swallowtail catastrophe. *Swallow's Tail and Cellos – The Catastrophes Series* – Dalí's final painting, finished in 1983 – is an almost purely mathematical image, with graphs corresponding to the equations for several of Thom's classification of catastrophes. There are also intriguing integral signs that look very similar to the holes that are carved into a cello. Indeed, a ghostly cello floats mysteriously in the top left of the painting. Two of the integral signs are put together as if in homage to Dalí's moustache.

Dalí didn't have the scientific or mathematical training to understand the deep insights that Thom's classification represented, but he was certainly sensitive to both the aesthetic nature of the work and the potential for mathematics to be the key to explaining violent physical phenomena. Throughout his life, Dalí constantly explored the fundamentals of science that have given rise to the world we inhabit: DNA, quantum physics, relativity and mathematics. When he died in 1989, the books on his bedside table included several by scientists such as Erwin Schrödinger and Stephen Hawking. He was wrestling with the mathematical blueprints of the universe right till the end.

As he once said: 'Thinkers and literati can't give me anything. Scientists give me everything, even the immortality of the soul.'

Living inside a dodecahedron

While painters from Leonardo to Dalí were restricted to delineating the Platonic solids on two-dimensional canvas, with all the challenges that this represents, architects are free to build actual versions of these shapes, if they so choose, in our three-dimensional environment.

Of the five shapes that are documented in Euclid's *Elements*, only the cube has proved a successful blueprint for architects. To my knowledge, the ancient world never used the tetrahedron as a shape for a triangular-based pyramid. One could argue that the Egyptian square-based pyramids look like half an octahedron sitting in the desert, hinting that there might be another shape embedded in the sands which completes the symmetry. But pyramids did not really catch on as living spaces. The cube and its cuboid relatives are the shapes that we saw Palladio or Le Corbusier putting together to create their iconic buildings.

There is a reason for this. Buildings are made up of rooms. Cities are made up of buildings. These different units need to maximise the space available. So architects are drawn to shapes that mesh together to fill space effectively. Of the five Platonic solids, only the cube can be used to fill space with no gaps. We can take elongated versions of the octahedron or tetrahedron and use these to fill space if we sacrifice complete symmetry, but the impracticality of living in an upside down pyramid means that these shapes are not going to make efficient building blocks for the architect.

But not everyone has been put off by the unwieldy nature of these more challenging shapes. If one is prepared to mix shapes and branch out to include some of the Archimedean solids as building blocks, then one has much more flexibility in filling space efficiently. The Israeli architect Zvi Hecker built an extremely striking synagogue in a military facility in the Negev Desert, which consists of a space-packing agglomeration of truncated tetrahedrons, truncated octahedrons and cuboctahedrons.

But Hecker was always particularly taken by the possibility of building with a pure dodecahedron as a blueprint. This isn't as unnatural as it might at first appear. Although dodecahedrons can't be put together to fill space, there is a distortion of the dodecahedron that does, and it can be found in nature.

Pyrite, or fool's gold, is an iron sulphide crystal that frequently can be found alongside copper. It predominantly takes the form of cubic crystals, but there is a second configuration, made from 12 pentagonal faces. The pentagons are not symmetrical, but by sacrificing the symmetry the shapes fill space. It's thought that the discovery of these crystals in Roman times might have provided the inspiration for artisans to cast in bronze some of the first regular dodecahedrons. They might have experimented with these crystals and found that you could actually level off the sides of a pyrite crystal to make each face a perfect pentagon. Hecker himself had seen how irregular versions of the shape often appeared in dry-stone walls.

Hecker wanted to take perfect dodecahedrons and to experiment with the sort of buildings they might make. One of the attractions of the shape is the way it is full of golden ratios, a proportion that all artists and architects seem drawn to, as we have seen with Le Corbusier's obsession with it. Hecker was keen to find a new way to follow in this great tradition.

The dodecahedron at least provides level floors and ceilings, allowing it to be stacked. But Hecker believed that the rooms inside a dodecahedral building would benefit from having some sides made up from the external shape. Hecker understood that because the shape of a dodecahedron tends towards a sphere, it would have a smaller surface area than a cubic room of the same size with conventional vertical walls, and hence would require less material for its construction. He also felt that a room with the sloping walls of a dodecahedron would feel more spacious and harmonious than a cubic room.

When Hecker was asked in 1971 to design a housing complex in Ramot, just outside Jerusalem, he decided this was his chance to put his passion for dodecahedrons into practice. The complex was in one of Israel's early illegal settlements in the occupied West Bank intended to shield Jerusalem. The area was not easy terrain for building, but this was one reason that Hecker liked the pentagon as a footprint. The fact that it didn't easily fit together in a regular tiling meant he could potentially be a bit more flexible in how he accommodated his plan to the lie of the land.

Work on the commission unfortunately coincided with the Yom Kippur War of 1973. Hecker was serving in the Israeli army, far from his office and cardboard models. But in between fighting, he began to make freehand sketches of how the dodecahedrons might be arranged to meet the project's specifications as well as his own aesthetic predilections.

Once the war was over, Hecker had the chance to turn his sketches into physical buildings. In the end, he had to resort to the trusty cube to help him realise the project. Although each apartment was in the shape of a dodecahedron, the only way to put them together was to arrange the centre of each apartment in a classic cubic structure.

But there were some advantages to the gaps that necessarily appeared between the apartments. The complex was to be occupied by Ultra-Orthodox families, and one of the requirements was that each apartment had an outside area that could be used for the week-long religious holiday of Sukkot. This is a time in the early autumn when Jewish families eat outside in temporary accommodation open to the sky in order to commemorate the Exodus from Egypt. It turned out that the gaps between the dodecahedrons provided the perfect space in which to build these *sukkahs*.

When I first saw the development during my time as a postdoc at the Hebrew University in Jerusalem, it took my breath away. The final result is extraordinary. It has been celebrated as one of the most bizarre buildings ever built. The fact that it still looks so outlandish compared to anything built since is testament to the dodecahedron's unusual nature when it comes to architecture.

Some architects, though, have started to experiment with buildings inspired by shapes in hyperspace. Dalí had successfully wrestled with the challenge of representing a four-dimensional cube on a two-dimensional canvas by cleverly unwrapping it into three dimensions and painting the resulting net. Given that architects have the luxury of making things in three dimensions without having to flatten them further, there is a chance to build structures that give a glimpse of these four-dimensional shapes beyond our physical universe.

One such project can be found in Paris: the Grande Arche. A conventional shadow is a two-dimensional image created by a light shining on a three-dimensional shape. As you emerge from the Métro into the financial district at La Défense, you are greeted by what can be interpreted as the three-dimensional shadow of a huge four-dimensional cube. It was designed by the Danish architect Johan-Otto von Spreckelsen, who used one of the tricks developed by Renaissance artists to depict three-dimensional cubes on two-dimensional surfaces. They would draw a square inside a larger square and by joining up the corners they would create the illusion of three dimensions. The image is the shadow that is created by shining a light above a wire cube frame.

Spreckelsen used the same idea at La Défense to represent what a shadow of a four-dimensional cube would look like. He has built a smaller cube nestled inside a larger one, and then joined up their corners to create this hypercube with its 16 vertices.

What of other symmetrical shapes out there in four dimensions? Mathematicians have discovered that it is possible to make analogues of the other Platonic solids in the same way that the hypercube extends the idea of the cube into higher dimensions. But rather intriguingly the fourth dimension allows for a sixth shape to be built. By piecing together 24 octahedral faces (remember that faces in these four-dimensional shapes are three-dimensional solids), you can make something called a '24-cell' or, as it is sometimes referred to, an 'octaplex'. If we push beyond four dimensions, then three of these shapes become impossible to extend, and we are left with just three Platonic solids: the tetrahedron, the cube and the octahedron. Hyperspace versions of the icosahedron and dodecahedron become impossible to build in five dimensions and beyond. We are still waiting for the artist or the architect to use these exotic shapes as blueprints for inspiration. But perhaps it will be the choreographers who find a way to let us experience these shapes.

Laban revisited

After Laban escaped Germany to Paris, he eventually made his way to England in 1938. Here he joined up with two of his former students at Dartington House in Devon, from where he continued his exploration of the importance of geometric form to dance. His time in England led eventually to the establishment of the Laban Art of Movement Guild, which today forms part of the Trinity Laban Conservatoire of Dance and Music. A whole generation of dancers has benefited from Laban's belief in the importance of the Platonic solids for how the human body moves in space.

One of the things that struck Laban from a very early point in his career was the observation that dancers, regardless of background, would emphasise the same points in space. The movements that they made between these points also seemed to have a commonality. He kept seeing the same patterns mapped out. He wrote how 'people, in spite of their differences of race and civilisation, had something in common in their movement patterns'. It was by mapping out these points that Laban discovered the importance of these Platonic shapes to the way people moved, especially the shape of the icosahedron.

Allow the body to move and, regardless of cultural background, it will map out the icosahedron. The universality of the underlying mathematics, Laban believed, gave rise to a theory of movement that transcended cultural boundaries. One of the interesting innovations that Laban introduced was the idea of dance scales. These were passages of movement around the corners of the geometric shapes that dancers were encouraged to imagine encasing their bodies.

It is an interesting mathematical challenge to discover if there is a way to visit all the corners of a shape, one after the other, by passing along its edges. Such a journey around a shape is called a 'Hamiltonian path', after the Irish mathematician William Hamilton. If you can return to the point you started from, then the path is called a 'Hamiltonian cycle'. He first proposed the idea in a talk he gave to the British Association for the Advancement of Science in Dublin in 1857, when he challenged his audience to find the path around a dodecahedron, referring to the puzzle as the 'Icosian game'. As with many mathematical ideas, however, the earliest discoverer was not given due credit. It appears that the English mathematician Thomas Kirkman had posed the same question for different shapes in a paper he had submitted to the Royal Society two years earlier.

Hamilton's friend John Graves thought the challenge would make a great toy and put Hamilton in contact with a toy maker in London who manufactured a version of the game and distributed it across Europe. In the game, the original three-dimensional dodecahedron was flattened into a two-dimensional network marked on a pegboard, which was traversed by winding a thread around pegs at each of the nodes until they had all been visited once. To make it all a little bit more appealing for children, the game was called Around the World and the pegs were labelled with the names of cities that you had to try and visit. Even so, the game was a flop. Finding the path for a flattened dodecahedron was just too simple.

Other shapes, however, are not so simple. For example, the Herschel graph – named after Hamilton's friend, the astronomer Alexander Herschel – is the smallest two-dimensional network of a polyhedron for which there is no Hamiltonian cycle. There is a path that visits all the nodes, but you end up stuck, unable to get back to the beginning and complete the cyclical journey.

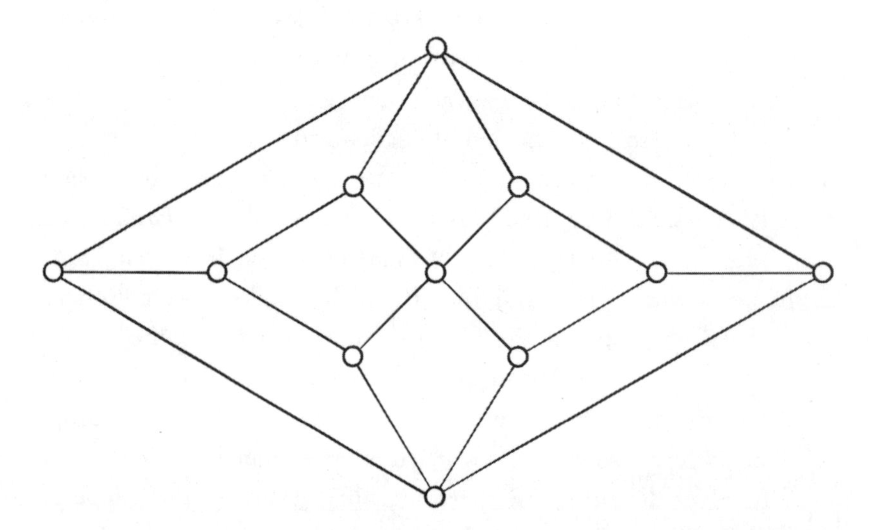

A Hamiltonian path, in contrast to a Hamiltonian cycle, is the name for such a path that visits every node without the need to come full circle and return to your starting point. It turns out that there are some networks that don't even have Hamiltonian paths. One interesting example was constructed by Bill Tutte in 1946, which he cooked up following his wartime activities cracking codes at Bletchley Park. This was a shape with the extra condition that only three edges met at each node. For such graphs, it had been believed that it would always be possible to draw a Hamiltonian path. If this were true, then it would have provided a simple proof of the four-colour map problem: a challenge to show that a map of the world needed no more than four colours to ensure that any countries with a common border were always differently coloured. The discovery that there was a shape without a Hamiltonian path scuppered this approach to the four-colour problem, which was eventually one of the first proofs that required a computer to confirm it.

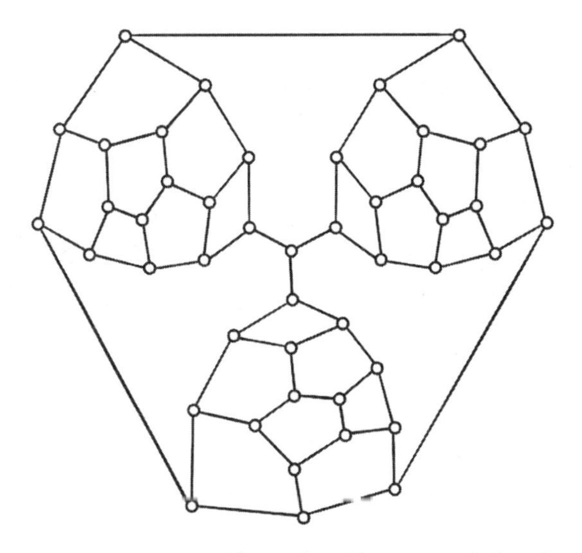

But if you restrict yourself to the Platonic solids, then there is always a cycle around the points of each shape. And these are what Laban used to make his dance scales. The idea was to provide a sequence of moves that could be used as the ingredients for more complex dances – much like the way that musical scales provide a performer with a shortcut to playing passages in a composition. They act a bit like words, so that rather than reading each letter individually you read a collection of letters in one go. Laban hoped his dance scales based on these mathematical shapes would play the same role in choreography.

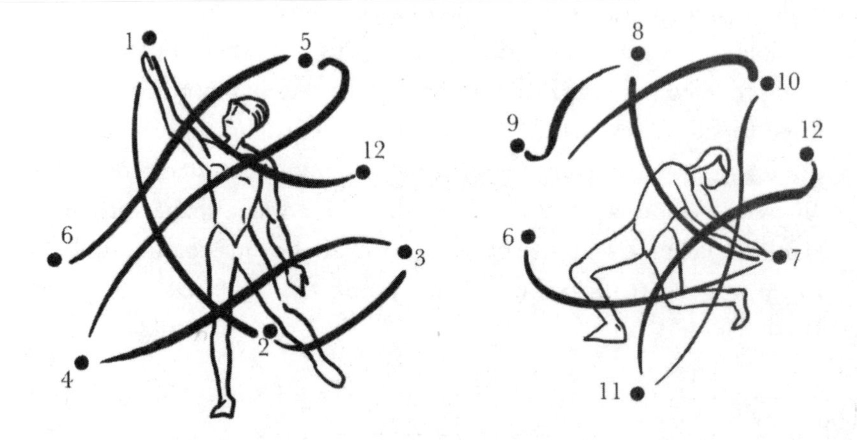

The icosahedron was always the shape that Laban believed was most attuned to the way a dancer moves, and it gave rise to different scales according to the different paths around the points. The 'A scale' and the 'B scale' mapped out the vertices of the shape by sweeping from the corners of one face to the corners of the opposite face in a series of moves called 'volutes'. The 'axis scale' and 'girdle scale' alter the emphasis of the movement to reflect axes passing through the icosahedron. But all the scales are examples of different Hamiltonian cycles that you can map around this three-dimensional shape.

One interesting aspect of these scales reveals itself if you imagine mapping out their paths with a rope. Sometimes doing this just gives rise to a loop, but quite often it results in a form of knot called a 'trefoil knot'. It seems that Laban was aware of this, and enjoyed the effect, because his notebooks contain many sketches of this trefoil pattern.

Laban's work has had a lasting effect on modern dance, helping to create a common language that allows dancers to communicate across cultural and linguistic divides. It is said that he did for dance what Stanislavski's systematic 'method of physical action' did for acting. For Laban, the universal nature of the Platonic solids meant they resonate with dancers from different backgrounds. He never felt like he created or invented his style of choreography, rather he talked of its discovery as if it existed outside the human realm, waiting for us to find it. It is

the universality of mathematics that Laban tapped into in order to create a common language of dance which would unite humanity in a way that his Olympic project had once hoped to achieve.

The Platonic solids that inspired Laban's choreography are just the opening salvo in humanity's exploration of one of the most pervasive concepts in mathematics, art and nature: symmetry, the focus of our next blueprint.

Blueprint Seven
Symmetry

It was after attending a concert at the Salle Pleyel in Paris in 1930 that the poet Raymond Queneau began to think about the potential of mathematics to inspire not only musicians but also writers. He'd just sat listening to Bach's final work, *The Art of Fugue*, with his friend Michel Leiris. The piece takes a simple theme and over 14 movements inverts, reverses, transposes and stretches it in ways that make it feel as if one is revolving a shape and seeing it exist in multiple different dimensions. Bach loved using ideas of symmetry where a shape is transformed and yet retains a connection to the original image.

The power of these mathematical tools of symmetry and transformation to create such infinite variety is one that music had exploited even before Bach turned his creative skills to articulating mathematics through sound. Queneau emerged from that concert convinced that writers were missing a trick. What if the same structures that had been inspiring musical innovation for centuries could provide blueprints for new writing?

His first experiments were with the power of number to encode ideas. The number 91 became an important structural piece of arithmetic in his first novel, *Le Chiendent* (known in English as *The Bark-Tree*), published in 1933. The text is divided into seven chapters consisting of 13 parts, giving a total of $7 \times 13 = 91$ parts. At one point, the characters in the novel carry round a door that they try to place in 91 different locations. For Queneau this number was not arbitrary. Rather it was, for him, like a signature encoding important components of his personality.

The number 13, for example, was significant 'because it has renounced happiness. As for 7, I took it, and still take it as the numerical image of myself, since my surname and my 2 first names are each composed of 7 letters and that I was born a 21 (7 x 3)' (Queneau was born on 21 February 1903).

The obsession with a numerological encoding of the author's name smacks of Bach's own obsession with the number 14 which was the total you got by translating Bach into numbers using gematria, the Hebrew practice of assigning numerical values to words. B=2, A=1, C=3, H=8, making 2+1+3+8=14. He insisted on weaving 14s into his music everywhere. That was the reason there were 14 fugues in the concert that Queneau had attended. The 14th fugue was in fact unfinished and ended on four notes that, in their German names, spelled out BACH.

As entertaining as Queneau's novel *Le Chiendent* is, I felt when I read it that the numerology was running weirdly tangentially to the actual narrative. It seemed arbitrary and not integral to the story, as if it was really there for Queneau's personal pleasure rather than for the reader's: 'I like my characters' entrances and exits to be very precise. That's how I work. I hope it isn't obvious. It would be terrible if it were obvious.' Frankly no one would have noticed the numbers had Queneau not referred to them later in an article he wrote about structure in *Le Chiendent*. Queneau had hoped the numbers would act like the constraints that rhyme and metre impose on a poet. 'A novel is a little like a sonnet, though it is much more complicated. I believe in things being highly constructed.' Queneau ultimately felt that this experiment in *Le Chiendent* was a failure in embodying structure in narrative.

Just as Bach went beyond embedding simple numerology in his music and employed more interesting mathematical structures to frame his compositions, Queneau too experimented with more subtle mathematical blueprints. *Le Chiendent* already had a structural symmetry embedded inside it, as it starts and ends with the same lines of text creating a loop, an idea that Bach had used in the Goldberg Variations.

It's possible that Queneau's loop might have been the inspiration for Joyce's in *Finnegans Wake*.

Queneau's first truly innovative application of mathematical structures to literature took its lead from that concert he'd attended in 1930. *Exercises in Style*, published in 1947, begins with a very simple story of only 10 sentences and then tells the same tale in 99 different ways. Each retelling is meant to represent a structural variation of the original story which could be applied to any narrative regardless of its content. Some of them took their lead directly from Bach. The fifth story is styled 'Retrograde', which in music means playing the theme backwards. This was a trick Bach liked to employ, where a theme would be accompanied by the same theme played backwards as in the example of the crab canon. For Queneau, the idea translated into telling a story from the end to the beginning. Others were more literary challenges. Telling the story in a sonnet or haiku. Translating the story into slang or creating a version written as an operatic libretto. But Queneau also started to experiment with things of a more mathematical character. One variation was even called 'Mathematical' or 'Géométrique' and involved rewriting the story as a mathematical formula. In a series of variations called 'Permutations' Queneau explores what happens to the story if you start permuting the words or letters around in a fixed manner. Same words, same letters, different order. ('Permuting' is the technical word mathematicians use for the process of rearranging things in a new order.)

This idea of permuting the components of a structure is one that Queneau encountered during his early years as a student, when he had aspired to be a mathematician, not a poet or writer. Already, by the time he was 13, Queneau was obsessed by mathematics, having discovered algebra and logic. Aged 17, he was writing in his diary: 'I went with Leroux to the museum. I am furiously studying mathematics.' Although he enrolled to study for a philosophy degree at the Sorbonne, he signed up for every course that had anything to do with mathematics and logic. He was reading papers by Einstein on relativity, Cantor on different sizes of infinity and Poincaré on chaos theory.

Probably his mathematical distractions were responsible for him failing his philosophy course, at which point he decided to jump ship to the faculty of science, where he majored in mathematics. Unfortunately, he ended up falling between the two, failing both maths and philosophy. Many years later, he commented on the collapse of his university career in mathematics:

> Where I made my mistake was in believing that I could fill the gaps . . . as I became perfectly aware when I enrolled in my first year of mathematics. After failing two or three exams I understood that I would never pass. For example, for me mechanics was opaque. And so were the conics (the delight, the *non plus ultra* of mathematics for specialists . . .).

But he never lost his love for mathematics. Although he might be remembered for his literary work, he also has a set of numbers named after him, which arose out of his interest in mathematical structures as a blueprint for poetry.

The Queneau numbers

Poetry inevitably has a certain affinity to mathematics because imposed structure is often what distinguishes poetry from simple prose. The pattern of poetry allows the mind to make connections and also to memorise more easily the text, especially important in an oral tradition before writing allowed easy recording of ideas. In India, mathematics was recorded in the form of poetry precisely because poetic structure allowed formulas to be more easily recalled.

Over the centuries, many different poetic forms have been invented and experimented with – ballads, sonnets, haikus, and so on. Often these restrict the number of syllables used in each line or demand a particular rhythmical structure. The end of each line might follow a chosen pattern of words that rhyme. A ballad, for example, often has a simple ABAB shape. In classical Persian poetry the *rubai* is a four-line poem with an AABA rhyming pattern. It was a form that the astronomer and mathematician Omar Khayyam enjoyed using in his poetry. *The Rubaiyat of Omar Khayyam*, the name given by Edward FitzGerald to his translation from Persian of a selection of Khayyam's quatrains, uses this pattern but goes one step further, where it is cunningly interwoven across stanzas. For example, if the first stanza of four lines has a pattern of rhymes AABA, then in the next stanza the third line's rhyme, B, is picked up and given prominence: BBCB. Each subsequent stanza in turn then picks up the rhyme introduced in the third line of the previous stanza: CCDC. If you are really clever, you can bring the thing full circle, finishing the poem by coming back to the rhyme that you first started with: XXAX.

Queneau became fascinated with one poetic form in particular, which was invented in southern France in the twelfth century by troubadour Arnaut Daniel in his lyric poem 'Lo ferm voler'. The poem has six stanzas of six lines each followed by a final stanza of three lines. Here is a translation of the first two stanzas of his poem:

The firm desire that enters
my heart no beak can tear out, no nail
of the slanderer, who speaks his dirt and loses his soul.
And since I dare not beat him with branch or rod,
then in some secret place, at least, where I'll have no uncle
I'll have my joy of joy, in a garden or a chamber.

When I am reminded of the chamber
where I know, and this hurts me, no man enters–
no, they're all more on guard than brother or uncle–
there's no part of my body that does not tremble, even my nail,
as the child shakes before the rod,
I am that afraid I won't be hers enough, with all my soul.

Known as a *sestina*, poets from Dante to Petrarch, Kipling to Ezra Pound, have experimented with poems in this form. Stephen Fry described it, in his book *The Ode Less Travelled*, as 'a bitch to explain but a joy to make'.

The curious feature of the sestina is that in the six stanzas made up of six lines, rather than having a regular pattern of words that might rhyme with each other, the last word of each line is used again in the next stanza but in a different order.

Suppose the first stanza had the following six words ending the lines: *one, two, three, four, five, six*. Then, in the next stanza, these words would occur again at the ends of the lines, but in a new order: *six, one, five, two, four, three*. One way to think of this is to put the words on the corners of a hexagon, but not in the most obvious way. Instead, a path inside the hexagon marks out for you the order in which the words appear in the first stanza.

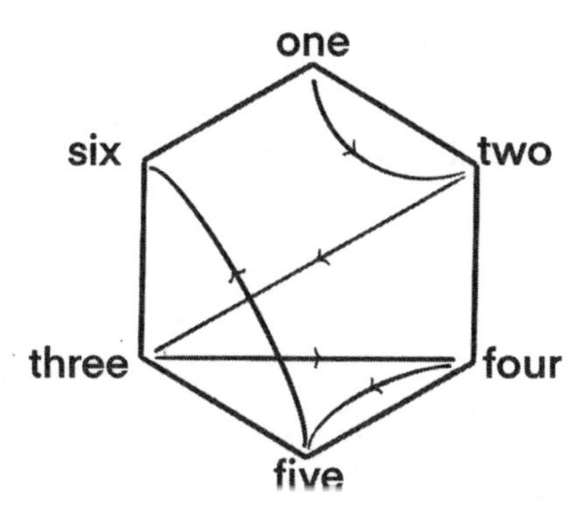

The clever trick is that to get the second stanza, you rotate the hexagon by a sixth of a turn, so the words move but you keep the path fixed so it doesn't rotate. So *one* moves to *two*'s spot, *two* to *four*'s and so on. Now use the path to read off the new order of words that appear in the next stanza. In each new stanza you turn the hexagon by a sixth of a turn, and this gives a new order of the end words. But notice that after six turns – six stanzas – the next turn would return the words to their original order. So the poetic form encodes the symmetries of a hexagon inside its structure. Here are the six rearrangements of the final words that the hexagon creates:

Stanza	1	2	3	4	5	6
	one	*six*	*three*	*five*	*four*	*two*
	two	*one*	*six*	*three*	*five*	*four*
	three	*five*	*four*	*two*	*one*	*six*
	four	*two*	*one*	*six*	*three*	*five*
	five	*four*	*two*	*one*	*six*	*three*
	six	*three*	*five*	*four*	*two*	*one*

There are 6×5×4×3×2×1=720 different possible permutations of the final words (or what mathematicians call '6 factorial', written '6!') but the symmetries of the hexagon severely restrict how things can be permuted because the rotation ensures that two points which are next to each other always remain adjacent. Notice that the first two lines of each stanza end with words that are next to each other in the hexagon. Or that lines 2 and 3 must end with words that are opposite each other in the hexagon. Randomness would tap into exploring all 720 possibilities. Symmetry cuts that down to six, massively restricting what is possible.

It appears that troubadour Arnaut Daniel was completely oblivious to the symmetry of the hexagon underlying his process. In addition to being a performer, Daniel was also an inveterate gambler, and it may be that his rule for how to permute the words came from a different shape with six sides: the dice.

The way the pips are arranged on a dice are such that the opposite faces add up to seven. The 1 is opposite the 6; the 2 opposite the 5; the 3 opposite the 4. This was a clever way of evening out any unintended biases that might have occurred with a cube that wasn't perfectly symmetrical. But this pairing of faces is one explanation for why the second stanza was chosen as it was: *six, one, five, two, four, three.*

Another way to get this order is to reverse the last three numbers – *six, five, four* – and then insert them, alternately before each of the first three numbers. But now you can keep doing this. Reverse the last three words of the previous stanza: so *two, four, three* becomes *three, two, four*. And insert these alternately before each of the first three numbers – *three, six, four, one, two, five* – and you've got the next stanza.

Queneau became fascinated by this poetic form because of its mathematical structure. He noticed that there is a rather beautiful way to interpret visually the rearrangement of the line-ending words from the first stanza to the second – by following them as they move round, you create a spiral. The third stanza would emerge by following the spiral again around the second stanza. So the sestina can be generated in two ways: using a spiral or the symmetries of a hexagon. And then the mathematical side of Queneau's character bubbled up: he began to wonder whether there was anything special about the number 6.

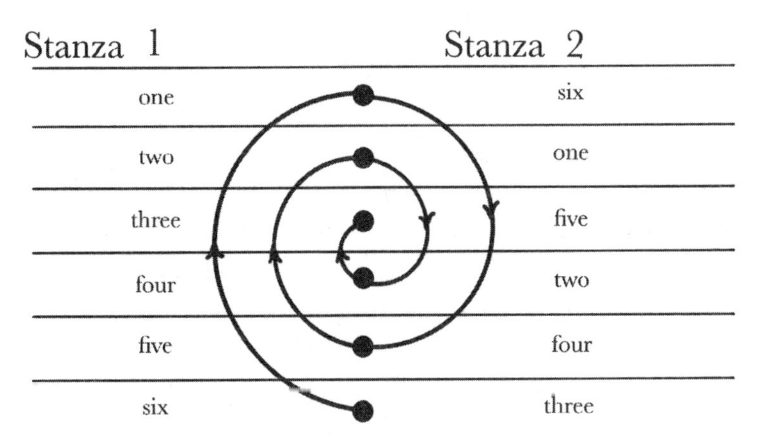

If you have a poem with seven lines and you keep rearranging the lines using a spiral, can you realise the same pattern by putting the seven last words on a seven-sided figure and rotating the shape like we did for the hexagon? It turns out you can't. The heptagon comes full circle after seven rotations but the spiral permuting the seven lines repeats after four repetitions; and not only that, the fifth line never moves. The words at the ends of lines 1, 7, 4 and 2 are arranged around a square which rotates; lines 3 and 6 flip; and line 5 is fixed. In contrast, a poem with five lines can be created both by a spiral and a pentagon. By putting the words on the points of a pentagon and rotating the shape you can reproduce the effect of the spiral. After five rotations of the pentagon or five iterations of the spiral the lines return to the original order and not before.

Queneau started experimenting and found by rote calculation that 31 numbers in the first 100 could give rise to satisfying poetic forms (at least from a mathematical perspective) whereby, for a poem of N lines, the line-ending words are fully rotated both by the spiral and the corresponding shape with N corners. The original sestina was sometimes known as a sextine, so these new forms were named 'quenines' in honour of Queneau, and the numbers for which the spiral works are now called 'Queneau numbers'.

Queneau used trial and error to tease out the first 31 Queneau numbers, but mathematicians are always after a general rule which would characterise all Queneau numbers. The answer was partially provided by Monique Berger, who was a student of Queneau's friend Jacques Roubaud, and it has a lot to do with the numbers in our first blueprint: the primes. If both N and 2N+1 are prime numbers then a poem with N lines will be an example of a quenine. For example, 11 is a prime and so is (2×11+1)=23, which means you could write a quenine with eleven 11-line stanzas. These special primes N where 2N+1 is also prime had actually been important in earlier work on Fermat's Last Theorem, and they are named after the French mathematician Sophie Germain. Born in 1776, she had to assume a male pseudonym in her correspondence with other mathematicians in order for people to take her ideas seriously. It is still unknown whether there are infinitely many Germain primes.

It turns out that, in addition to the Germain primes, there are also some other numbers that generate quenines. A complete characterisation of Queneau numbers was finally provided by Jean-Guillaume Dumas in 2008 and is related to deep questions of number theory. Despite this characterisation it has still not yet been determined whether there are infinitely many Queneau numbers or not.

All this mathematics may have left your brain fried, but Fry offers some encouraging words:

If like me, formulae with big Greek letters in them mean next to nothing, you will be as baffled by it as I am, but you might like, as I do, the idea that even something as ethereal, soulful and personal as a poem can be described by numbers . . .

Surviving with symmetry

One of the interesting things about the sestina is that the brain, as it engages with a poem in the form, starts to pick up the evidence of a pattern between stanzas. The brain has evolved to be hypersensitive to these hidden patterns. It can sense if not articulate the symmetry of the hexagon underpinning the way the words are moving around the poem.

Symmetry is used in nature very often to signal that something is worth taking notice of. In the chaos of the jungle, if you see something with symmetry then it is likely to be an animal hiding in the undergrowth. Either that animal is going to eat you, or you can eat it. Recall Blake's famous poem:

Tyger Tyger, burning bright,
In the forests of the night;
What immortal hand or eye,
Could frame thy fearful symmetry?

Many animals rely on symmetry to help them navigate the chaos of the natural world. For example, the eyesight of a bee is extremely limited. As it flies through the air in search of food, it has to find some way to make sense of the onslaught of images it is bombarded with. Evolution has tuned the bee to recognise shapes full of symmetry because these are likely to be flowers, where it will find the sustenance that will keep it alive. The flower is equally dependent on the bee for its survival. It has evolved into a symmetrical form in the hope of attracting the bee. Symmetry acts like a language communicating information across the natural world.

By recognising symmetry, there is a better chance of survival. But it has led to a brain that can't help but search out symmetry in whatever it encounters. The pattern in the sestina is intriguing for the brain because it recognises the recurrence of the words, but their new order then challenges the brain to try to predict where they will end up next. The brain attempts to reverse-engineer the sestina to understand the blueprint from which it was constructed. It is the same sensation that the brain experiences when it is listening to a piece of music such as *The Art of Fugue*, which inspired Queneau. The reason the brain fires up when it is exposed to symmetry in poetry, music or art is because it has evolved to be sensitive to the symmetry that it encounters everywhere as we navigate our way through life.

Symmetry is probably the most ubiquitous blueprint that we find underpinning the different structures in our universe. Nature enjoys hiding mysterious symmetries at the heart of many parts of the natural world – fundamental physics, biology and chemistry all depend on a plethora of symmetrical objects. The six-sided symmetry of the snowflake, the eight-sided symmetry of the medusa, the simple reflective symmetry of the human face are some of the obvious manifestations of how much nature loves symmetry. But symmetry in the natural world isn't simply a thing of beauty; it serves a function for each object that displays it. The honeycomb built by the bee is made up of hexagons only because this six-sided figure is perfectly adapted to packing things efficiently and minimising the costly effort of producing the wax needed to create it.

As we observed in the blueprint dedicated to the circle, the spherical symmetry of a bubble is related to the fact that this shape minimises the energy needed to maintain the bubble. Nature is very lazy. The sphere is also a very strong shape, which is why we use it for divers' helmets and submersibles.

Symmetry has been at the heart of our theory of the natural world ever since Plato associated five symmetrical objects with the building blocks of nature. Although we no longer believe in a universe built from Platonic solids, symmetry is still central to our understanding of the cosmos. Recent theories in cosmology even indicate that the universe has a dodecahedral shape, an idea that Plato struck on more than 2,000 years ago. But it is not only on the cosmic scale that one finds symmetry. Symmetry is also key to the molecular world. The crystalline structure of many molecules reveals a whole host of different symmetrical possibilities. The structure of diamond, for example, is built around the symmetry of the tetrahedron, which is the key to the crystal's remarkable strength.

Molecular biology also exploits symmetry. The protein case that surrounds a virus often exploits a symmetrical shape, such as the icosahedral shape of the herpes or Covid virus that we encountered in the blueprint exploring the Platonic solids. The DNA enclosed inside the protein case essentially acts like a program to recreate copies of the virus. But often the length of the DNA is very short, so the program has to be very efficient. The symmetry of a shape like the icosahedron provides a very simple program for constructing the whole of the object from a simple building block.

If one digs even deeper, one finds symmetry is also at the heart of the subatomic world. The menagerie of fundamental particles revealed by physicists' supercolliders only makes sense when you start to see them as facets of some strange higher-dimensional symmetrical shape. These symmetrical blueprints have led to the discovery of new particles because there were facets of these strange higher-dimensional shapes that didn't correspond to any of the particles we had previously seen. The symmetry guided us towards the pieces of the jigsaw that we'd missed. Plato's insight that the building blocks of the physical universe are symmetrical shapes has been borne out by modern particle physics.

In contrast to the symmetry we encounter at the molecular scale in the natural world, animals find it quite difficult to achieve symmetry. This is part of the reason why we are attracted to people with a lot of symmetry: if they can waste energy making themselves symmetrical, it is a mark of good genes and a healthy upbringing. Symmetry is an indication that someone will make a good mate. It is nature's way of declaring: 'My genes are so strong that I can waste time making myself symmetrical.' That's why the peacock dedicates such effort to creating a tail that serves no other purpose than showing off to a mate what a great partner they will be. Tests have revealed that chickens in a battery farm, where they have to work very hard to survive, create very unsymmetrical eggs. In contrast, the eggs of free-range chickens are much more symmetrical. Symmetry is a sign of good health both in nature and nurture.

Symmetry is a fundamental concept across the sciences. But as we have seen, it is also important to many parts of human culture. From architecture to music, from poetry to painting, symmetry underpins many of the structures used in the creative world. The evolutionary advantage that a brain sensitive to symmetry bestows means that this is a blueprint that we are drawn to time and again.

For Alexander Pushkin, Russia's greatest poet, there is no escaping the importance of the concept: 'Symmetry is a characteristic of the human mind,' he wrote in a letter to a friend. The French poet Paul Valéry also recognised that symmetry is at the heart of the way we perceive our surroundings: 'The universe is built on a plan the profound symmetry of which is somehow present in the inner structure of our intellect.' And it's perhaps not surprising that they are drawn to express their art through poetry. The rigid logic of its rhyming structure and its rhythmic patterns makes traditional poetry the literary form that most resonates with the mathematics of symmetry.

But symmetry is not a concept that is universally loved by creative artists. This is Thomas Mann describing a snowflake in his novel *The Magic Mountain*: 'the living principle shuddered at this perfect precision, found it deathly, the very marrow of death.' For some artists, the rigidity of perfect symmetry boxes one in, forces things to be too predictable. Many of them, for example dismiss Escher as second-rate because his obsession with things symmetrical cramped creativity, leaving no room for emotional expression.

But there is one art form above all others which successfully exploits the blueprint of symmetry for creative inspiration: music. Queneau experienced the power of symmetry as he listened to that concert performance of *The Art of Fugue*. From Bach to the composers of the twenty-first century, symmetry offers a dynamic rather than a static tool for navigating a journey through sound.

A modern-day Pythagoras

The idea of using the symmetry of a shape to chart alternative pathways through music was the inspiration for an extraordinary piece for solo cello created by the Greek composer Iannis Xenakis in 1965. As a young man, Xenakis had played an active role in the communist resistance during the Second World War first to the Axis occupation of Greece and then to British attempts to restore the Greek monarchy. This plunged the country into civil war, during which he miraculously survived being hit by shrapnel from a shell fired by a British tank, although it cost him the sight in one eye and left his face hugely disfigured. Fearing for his life after the authorities started rounding up former left-wing resistance fighters, he fled Greece on a false passport in 1947, arriving in Paris with no valid papers and no money.

His luck turned when Le Corbusier offered him a job in his studio. Xenakis had studied architecture and engineering before the war broke out; he had managed to resume his studies and graduate with a degree in civil engineering just before he escaped Greece. Le Corbusier's studio had a policy of engaging refugees who arrived in Paris after the war, and Xenakis was hired to work on the structural calculations for one of Le Corbusier's Unité buildings.

Xenakis soon graduated from mere calculator to architectural collaborator. Indeed, one of Le Corbusier's most striking buildings, the 1958 Philips Pavilion in Brussels, is now recognised to be principally the work of Xenakis, although Le Corbusier had rather ungenerously claimed the design as his own, aided by contributions from his Greek employee. What is interesting about the pavilion is that its geometry had also been the inspiration some years earlier for the creation of a piece of music by Xenakis.

Xenakis had been passionate about music ever since his days at school and had combined his engineering degree with courses in musical composition. In Paris, he was keen to continue his studies, and he attended classes by some of the leading French composers. His compositions were not well received by many of his teachers, and he was discouraged by their insistence that he go back and relearn the basics of counterpoint and harmony. But then a conversation with one of them changed his mind. That teacher was none other than Olivier Messiaen. When Xenakis asked Messiaen after a class one day whether he agreed with his fellow teachers that Xenakis should go back to basics, he got something of a shock.

Messiaen recalled the advice he gave: 'I did something horrible which I should do with no other student.' Usually Messiaen was a stickler for requiring his students to study harmony and counterpoint before they went on to compose their own work. But he could see that Xenakis was not an ordinary student. So he told Xenakis 'you are almost thirty, you have the good fortune of being Greek, of being an architect and having studied special mathematics. Take advantage of these things. Do them in your music.'

In 1954, following the completion of his studies with Messiaen, Xenakis composed an orchestral piece called *Metastaseis*. Its score looked more like a series of mathematical diagrams than music. The shapes closely resembled the hyperbolic surfaces that he would have encountered during his mathematical studies (and which we will explore further in our next blueprint). These were meant to instruct the string players how to perform certain glissandos in the piece. Their appearance, though, had remarkable resonances with the architectural plans for the Philips Pavilion that Xenakis produced some years later. Additionally, the second movement featured time durations controlled by the Fibonacci sequence, something that he would have seen used spatially in Le Corbusier's studio, in the shape of Modulor Man.

Xenakis was incensed when Le Corbusier refused to acknowledge him as the joint architect of the pavilion. He decided to write directly to the Philips company, who had sponsored the construction, to assert his authorship. 'It is I who entirely conceived the form and mathematical expression of the Philips Pavilion. I now demand very firmly that your press service mention my name in the architectural creation of the Pavilion.' But Le Corbusier could not countenance sharing credit. Although he had hundreds of architects working under him, he was the name people paid for.

The struggle that ensued led to Xenakis being dismissed from Le Corbusier's studio. Xenakis realised he had been sacked only when he returned from his summer holiday to find the locks of the office had been changed and he had not been sent the new keys. Secretly, Xenakis was quite happy to move on. It meant he could concentrate on his true passion: music.

Variations on a cube

Xenakis remained true to Messiaen's advice and continued to plunder his mathematical and architectural knowledge for musical inspiration. There is one piece in particular, which Xenakis composed in the year I was born, and which has obsessed me for years because it taps into my passion for the mathematics of symmetry and my amateur attempts to learn the cello. This is the work he created in 1965 for solo cello using the symmetries of a cube as a blueprint.

Called *Nomos Alpha*, the composition is divided into 24 sections, each of which is a bit like a stanza in one of Queneau's sestinas. The first section Xenakis visualised as a cube in which on each corner he placed one of eight musical 'textures', comprising pizzicatos, glissandos, tremolos and other techniques that the cello can perform. These corners are played in a fixed order corresponding to two tetrahedrons that can be embedded inside the cube. At the end of the section, the cube is rotated to rearrange the musical textures. The same path is mapped out in the new section, but the textures are now played in a new order. There is also a second cube that rotates from one section to the next but keeps track of the time spent on each corner and the dynamics across the piece.

The eight musical textures on the corners of the cube are like the six line-ending words permuted in a sestina. The path round the corners of the cube is fixed, but the rotation of the cube causes the textures to be played in a different permutation or order. As we have seen, the symmetry at the heart of the sestina is very simple. It is just rotations of a hexagon which return the words to their original order after six turns. The symmetries of a cube, though, are a bit more interesting.

There are 24 ways to rotate a cube from its starting position. Nine rotations of a quarter turn around an axis running through the centre of opposite faces. Six rotations of half a turn around an axis through the centre of opposite edges. And then eight rotations of a third of a turn around an axis through opposite corners. That makes 23 rotations, but we mathematicians add a 'zero rotation' – which is just to leave the cube where it is – to produce 24 rotations in all. Mathematicians call this the 'identity' symmetry and, like the revolutionary idea of including zero as a new number, it plays a useful role in the mathematical language of symmetry.

Since *Nomos Alpha* is divided into 24 sections, my immediate thought when I started exploring it was that there would be a section for each symmetry. However, as I dug deeper, I discovered that something much more interesting was going on.

The 24 sections of the piece are arranged into six groups of four. The first three sections in each group correspond to three symmetries of the cube. The fourth section is more fluid and imagines the cube morphing. What fascinated me was how Xenakis chose the symmetries corresponding to each section. He started with two seed symmetries, identified in his notation as 'D' and 'Q12'. D is a third of a turn through opposite corners. Q12 is a half turn through opposite edges. These control the permutation of the textures in the first two sections. But to get the symmetry for the third section, the cube is first rotated using symmetry D and then rotated again using symmetry Q12. The combined effect is a new arrangement that can also be obtained in one go with symmetry Q4, a rotation of a quarter turn around an axis through opposite faces. As the piece continues, skipping every fourth section which is more freeform, the symmetry for each new section is got by combining the symmetries from the preceding two sections. So, passing over the free-flowing fourth section, the symmetry corresponding to section five is determined by combining the symmetries of sections two and three.

This immediately reminded me of the way the Fibonacci numbers are defined by adding together the two previous ones: 1, 2, 3, 5, 8, 13, . . . These were numbers that Xenakis was encountering every day while working in Le Corbusier's studio, and could easily have been the inspiration for this choice.

While the Fibonacci numbers spiral off to infinity, these Fibonacci symmetries eventually return to the original two symmetries you started with and then repeat themselves. How quickly the symmetries repeat depends on the two symmetries with which you start the process. For the two symmetries that Xenakis chose to begin *Nomos Alpha*, it takes 18 rotations before the sequence repeats. This is the longest cycle of symmetries you can realise with a cube. If you choose two different seed symmetries then the sequence can be shorter, so it would be intriguing to know how much experimentation Xenakis did to discover this longest path. The 18 symmetries that you encounter on this Fibonacci journey are not unique. Some of them occur twice. But it takes 18 symmetries before the original two reappear and the entire journey repeats itself all over again.

section	1	2	3	4	5	6	7	8	9	10	11	12
rotation	D2	Q12	Q4	free	E2	Q8	Q2	free	E2	Q7	Q4	free

section	13	14	15	16	17	18	19	20	21	22	23	24
rotation	D2	Q3	Q11	free	L2	Q7	Q2	free	L2	Q8	Q11	free

Nomos Alpha is not an easy listen. But understanding how the symmetries of a cube control its structure gave me a way to engage with a piece which, on my first encounter with it, had felt very alien. It provided a conceptual map of the journey the cello makes which helped me navigate its complexities. Inspired by this, I teamed up again with the artist Simon Russell – whom I'd worked with to develop a visual journey through Messiaen's *Quartet for the End of Time* – to create an animation that you can watch alongside Xenakis's piece to help you comprehend it – or, if you know it, to hear it in a new way. I also hoped that it might assist performers trying to learn the piece.

The animation has proved very popular and you can watch it online (a link is provided in the Bibliography). I've even gone as far as playing it live at the Southbank Centre in London to accompany a performance of *Nomos Alpha* by the cellist Julie-Anne Manning. The animation consists of 150 micro-animations, corresponding to each corner being played in each of the 18 sections controlled by a symmetry of the cube together with the 6 free-form sections. Cueing each shift required me to follow the score like a hawk in order to synchronise each move with the cellist.

The performance became an actual duet when we encountered the last section, which is also the sixth one where we hear and see the cube morph. In the previous morphing cube sections, the bottom string of the cello has to be tuned down an octave. This is often solved in performance by having a second cello, with the string already retuned, which the performer can pick up and play.

The last section, however, requires the cellist to play two lines of music simultaneously. One line ascends into the high notes, while the second descends into the bass of the cello. Simply put, it is impossible for a single cellist to play this live. There are different ways to solve the challenge. Often a cellist has pre-recorded one of the lines, which is played from a speaker hidden beneath their chair while they perform the other line on stage.

However, for our performance at the Southbank, we found an alternative solution. When we reached the final section, I picked up the second cello and, as the animation unfolded, we played the final section together. It was a fun finale to the event. Later, I was keen to know how Simon and my visual mapping of the piece had affected Julie-Anne's performance. She told me in an email:

> The animation helped to bring to life the architecture of Xenakis's Nomos Alpha. Although I can't say I understand all of the mathematics, the structure and relationship between the sounds and the different apexes of the cube became more obvious. It's fun and interesting to perform the work whilst seeing the audience following along on their mathematical journey.

Perhaps the most intriguing thing to come out of this whole adventure was a new mathematical concept: the idea of a Fibonacci sequence of symmetries. It is not something that had ever been considered before, but Xenakis's novel contribution to mathematics sent me off on my own journey of research into symmetry. Xenakis had discovered that, for a cube, the maximum loop length of symmetries is 18, but what about for other shapes? Are there any patterns to the numbers? Is there a way to predict the lengths of the loops? Often mathematics inspires music, but here we see music inspiring new mathematical ideas.

Nomos Alpha was dedicated to three mathematicians who had contributed to the mathematical understanding of symmetry. One of them was the Frenchman Évariste Galois, who developed an algebraic language for symmetry called 'group theory'. It is a language I use every day as a research mathematician. It allows mathematicians to create and navigate symmetrical shapes that live beyond the physicality of our three-dimensional universe.

The grammar of symmetry

Killed in a duel in 1832 over love and politics at the age of 20, Évariste Galois is probably one of the most romantic characters in the history of my subject. Already, as a schoolboy, he had started to develop a new language that would allow us to perceive symmetry beyond the three-dimensional shapes of Plato and Archimedes. But when he presented his ideas to the French Academy of Sciences, they were dismissed or ignored as the ramblings of an untrained amateur. Partly this was a result of the young man's inability to articulate his new ideas.

His frustration at the mathematical establishment's failure to recognise his breakthrough led the young Galois to join a radical revolutionary movement that wished to overthrow the restored French monarchy. The authorities arrested him twice: once for threatening the life of the king in a toast during a dinner with his republican comrades, and again for wearing the uniform of the disbanded National Guard. While he languished in prison, he continued to work on his mathematical ideas. When he complained to the guards about ill treatment, he was thrown in solitary confinement. The other prisoners leapt to his defence: 'Galois in the dungeon! . . . Oh, the bastards! They have a grudge against our little scholar. This young Galois doesn't raise his voice, as you well know; he remains as cool as his mathematics when he talks to you.'

When he was eventually released, he became embroiled in an argument with a fellow revolutionary, possibly over a woman called Stephanie. Their quarrel ended in the duel that killed Galois. Aware that he was probably about to meet his destiny, he spent the whole night before the encounter trying to correct and explain the principles of the mathematical language that he had developed. Probably the fact that he stayed up all night doing mathematics was one of the reasons he was such a bad shot that fatal morning. But the papers that he left behind on his table addressed to his friend and fellow mathematician Auguste Chevalier are today regarded as some of the most important documents in mathematics. For they contained the secrets of a new way to think about symmetry.

Galois's motivation for developing his language was connected to the age-old problem of solving equations. At school, he had learnt about a formula which told you how to solve a quadratic equation. Given the description of an equation, there was a formula which would find you the numbers that solved it. This discovery goes back to work by the ancient Babylonians. Mathematicians in the sixteenth century had subsequently come up with formulas to solve cubic and quartic equations. But the next step, to quintic equations, was proving elusive. The reason was that a phase change happens at this point. There is no formula that will solve quintic equations. Proving this required Galois's new language of symmetry.

There are several key steps in understanding the mathematics of symmetry. Most people, when you say 'symmetry', think simply of left–right reflectional symmetry in a human face, or perhaps the rotations of a hexagon. But mathematicians started to understand that the concept was more ubiquitous than these simple examples. A symmetry of a structure is some rearrangement so that the new configuration resembles the original one. In the case of the cube that Xenakis used to frame *Nomos Alpha*, each symmetry is a way to pick up the cube, move it around, and place it back down again so that the cube looks like it did at the outset but the corners have been rearranged. As we have seen, there are 24 such symmetrical moves for a cube. And here is where we can start to see a way to disconnect symmetry from the physical form. Because, for example, we can take a pack of cards, shuffle the cards into a new order and regard this shuffle as a symmetry of the pack of cards.

Symmetry starts to become more about describing how the defining parameters of a structure might be rearranged to retain the configuration of the original structure. For example, if we put eight cards on the corners of a cube, then not every shuffle of the cards can be achieved by the cube's symmetries. There are constraints that the underlying structure puts on the possible rearrangements. Recall how the symmetries of the hexagon at the heart of the sestina forced the first two lines of each stanza to end with words that were adjacent to each other on the hexagon.

The great breakthrough that Galois made was to understand that it wasn't just about listing all the symmetrical moves that a structure might have but how these symmetries might interact with each other. As Xenakis realised in *Nomos Alpha*, if you have two symmetries of a cube, you can combine these to make a third symmetry.

This realisation implied that there was an underlying grammar, or algebra, which connected all the symmetries together. Recall how Xenakis took symmetry D, a third of a turn through opposite corners, then followed it with symmetry Q12, a half turn through opposite edges. The combined effect is a new arrangement that you can get in one go with symmetry Q4, a rotation of a quarter turn around an axis through opposite faces. We write this as an algebraic equation which describes how these symmetries relate to each other:

$$\mathrm{D} \circ \mathrm{Q12} = \mathrm{Q4}.$$

One of the interesting consequences of this approach is that it gave mathematicians a language to understand when two outwardly different-looking structures might nonetheless be manifestations of the same underlying symmetry. For example, the symmetries of a tetrahedron and the shuffles of four cards are actually identical symmetrical structures. We can name all the symmetries of each structure, and the way they interact will be the same whether they are describing the shape or the shuffle of the cards.

One way to see why these are the same is to place the four cards on the faces of the tetrahedron. Every shuffle of the four cards can be achieved by a symmetry of the shape. What is fascinating is that although the language for understanding the potential and limitations of symmetry did not fully appear until the end of the nineteenth century, when Galois's ideas were finally decoded, there is already evidence from centuries before of artists discovering new sorts of symmetries beyond simple reflections. The artists of the Islamic world, in particular, were masters of the visual art of symmetry.

Reading the Alhambra

One of my favourite places to visit as a connoisseur of symmetry is the Alhambra in Granada. Sitting like a jewel atop the hills of the Andalusian city, it is one of the most beautiful palaces built by the Moors in Spain. We've already talked about the fractal *muqarnas* that one can find in the corners of the Alhambra's rooms. What draws most visitors to the palace is the elaborate stucco tiling that adorn its walls.

Although each room provided artists with yet another canvas to express their creativity, the words of the Muslim Hadith, which interpret the Quran, have imposed some limitations on them. It is forbidden under strict Islamic law to depict any living creature. So, instead, artists have been forced to represent the majesty of creation through more geometric imagery. And that is what makes the Alhambra such a feast for a mathematician.

To the untrained eye, the different games that the artists of the Alhambra could play with the symmetrical pieces look unlimited. But thanks to Galois's language, a mathematician stares at these walls through very different glasses to the average visitor. For instead of the infinite complexity of the different tiles and colours, the mathematician sees just 17 different symmetrical games being played by the artists. What I find intriguing is that, even before mathematicians developed a language to articulate what is going on in these designs, the artists that decorated the palace for the Nasrid dynasty between the thirteenth and fourteenth centuries seem to have had an intuitive grasp of the structures underpinning their designs.

As you wander through the palace, the symmetrical designs change almost as if to reflect the different characteristics of the building. In the more formal parts of the palace that greet the visitor as they arrive for the first time, you find designs based on the tiling of the square. These are designs with rigid reflectional symmetry. But as you proceed into the more private and intimate parts of the palace, the symmetry changes. The harem, for example, is dominated by symmetries of triangles and hexagons: 3s and 6s, rather than the 2s and 4s of the earlier part of the palace. Reflective symmetries are now disrupted by the greater emphasis on rotational symmetry.

Although the decorative patterns on the walls around the palace might all look very different, it is intriguing how those with the same underlying symmetry find themselves in the same parts of the palace. The artists seem to have had an intuitive sensitivity to the fact that the symmetry was identical despite lacking the language yet to articulate it.

For example, one of my favourite designs in the Alhambra are these triangles that you can find tiling one of the walls in the harem. I always think that the curved edges give the illusion that they are shimmering in the Andalusian heat.

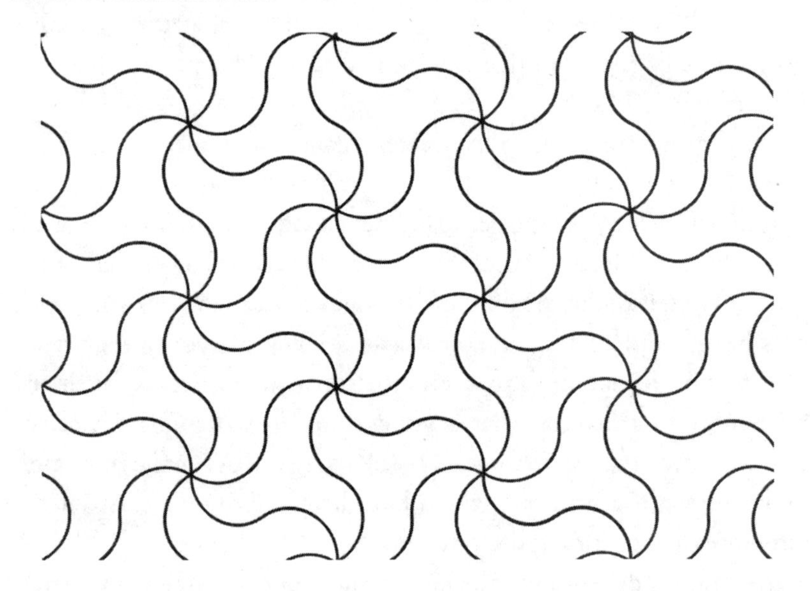

How does one use Galois's language to read the symmetries in this design? Note that it is the symmetry of the way the triangles are arranged on the whole wall that we are exploring, not the symmetry of the individual tile. A symmetry of the design will be a way to move all the tiles together so that after the move you can set them back down in their original outline. We are going to try to identify the range of different symmetries that this wall allows, called the group of symmetries of the wall.

The first thing to notice is that there are no lines of reflectional symmetry. The wave on each side of the triangle prevents you from simply being able to flip them over and lay them back down within the original outline of the tiles. But there are three distinct points around which you can rotate all the triangles together and set them back down in their original outline, symmetrical moves that preserve the underlying pattern. The first obvious symmetry is to keep the centre of a triangle fixed and then rotate all the tiles by a third of a turn around this central point. A second symmetry corresponds to rotating all the tiles by a sixth of a turn around the point where six triangles meet. A sixth of a turn around this point moves all the tiles so that they still fit inside the original outline. Finally, there is an interesting third symmetry, which involves fixing a point halfway along an edge and then spinning all the tiles through a half turn. This swaps the tiles that are joined along the edge. All the other tiles are also swapped around. These are the three basic symmetries of this wall.

But in the same part of the palace, you can find a tiling made up of strange Z pieces and six-pointed stars.

Although the star has reflectional symmetry, the fact that it is surrounded by the Z pieces destroys any reflectional symmetry in the overall design. This wall looks very different geometrically to that previous one, but once again there are three points in the design around which you can make rotations of a sixth, a third and a half turn.

So although the motifs featured on each wall are very different, mathematicians can identify the group of symmetries in each design as identical. The name that mathematicians have given this particular group of symmetries is '632'. Actually, these and other groups of symmetries were named originally by chemists, because the feature is important for describing the structures of crystals. But the names they used were so illogical that mathematicians frustrated at the chaos stepped in and devised a much clearer system of nomenclature.

The triangular tile with its shimmering sides was the inspiration for

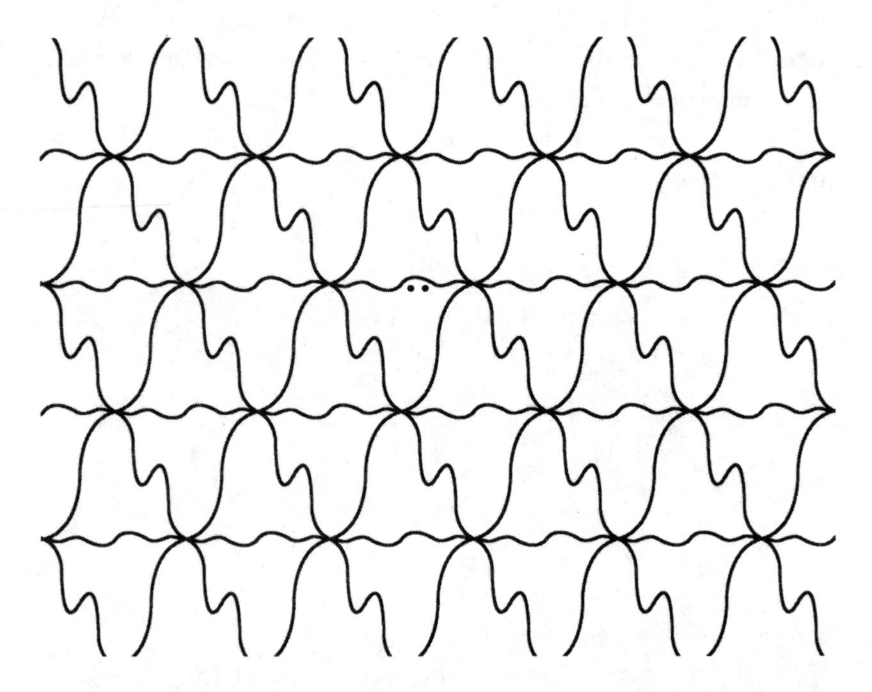

a new tile that I designed with artist Richard Rhys at the Pattern Foundry. The wave running down the side of the tile reminded me of a musical note. Richard and I wondered about varying the frequency of the wave on each side of the triangle to create a chord. If you compare the three notes on the sides of our tile they are in a 1 to 2 to 3 relationship. These are the harmonic frequencies discovered by Pythagoras. Our tile was an attempt to capture the idea of frozen music.

The symmetry underlying the tessellation of these tiles is called '2222', which indicates four distinct points with a half-turn symmetry. When we looked at the finished product, we suddenly saw this little ghost staring out at us. So we distorted one of the waves a little more in order to bring out the ghostlike quality of the tile. We even managed to make the tiles, thanks to a collaboration with one of the oldest tile manufacturers in England. So if you fancy tiling your bathroom with symmetry 2222, then get in touch. The Ghost Tiles come with a little surprise: each order contains a single tile that includes glow-in-the-dark eyes to reinforce the idea of the ghost.

The power of Galois's language is such that it allows us to go beyond the visual and perceive the abstract symmetry in designs like the Ghost Tiles or the tiles in the Alhambra. This abstraction of structure underlying the geometry is a bit like the idea of the abstract concept of number. Just as the number 3 captures the identity of a collection of three objects, whether they are three apples or three kangaroos, the naming of the group of symmetries 632 has abstracted the symmetrical identity shared by the two walls in the Alhambra. Galois has done for symmetry what the concept of number does for counting.

It was the language that Galois invented in the nineteenth century which finally revealed that there are only 17 different groups of symmetries that the arrangement of tiles on a two-dimensional wall can have. Any attempts by the medieval Moorish artists to conjure up a new wall of symmetry would have been doomed to failure. Whatever games were played, the wall would have had a group of symmetries identical to one of the 17 in the mathematicians' list. But it's striking that even without this language, Islamic artists created examples of all 17 structures in the palaces they built across Spain and beyond.

The symmetries in the Alhambra are realised in two dimensions. But one of the most exciting consequences of Galois's new language is being able to navigate symmetrical objects that exist beyond our three-dimensional world of cubes and spheres. This is what my own research concerns: discovering ways to tile palaces in dimensions beyond those of the Alhambra.

My mathematical poetry

Discovering new symmetries is one motivation for my work as a research mathematician but, just as for the artists of the Alhambra, beauty is another drive. As G. H. Hardy articulated in his book which inspired me to become a mathematician: 'Beauty is the first test: there is no permanent place in this world for ugly mathematics.' The aesthetic, emotional and creative side of mathematics has always been an important blueprint for my own journey in this world. The discovery that I am most proud of making is that of a new symmetrical structure which lives beyond our three-dimensional world.

I'm proud not just because of the scientific feat of uncovering a new symmetry for the first time, but because the unexpected properties that this structure embodies have an artistic, almost dramatic, side to them. The story of these surprising traits has the narrative quality that those proofs in Hardy's book first excited in me. When I entered the field as a young mathematician, the consensus was that symmetries behaved in a rather simple manner, one that was articulated in various conjectures by my mathematical forefathers. But my research led me to doubt that received opinion, and to believe that the world of symmetry might be connected to a very different area of mathematics: the number theory of elliptic curves.

One of the great remaining challenges in mathematics is the search to find solutions to an equation such as $y^2=x^3-x$, a quadratic equation in y equals a cubic equation in x. This is one example of an elliptic curve. The difficulty is finding whole number or fractional solutions to them. If you enter a value for x in this equation, then to be a solution y has to be the square root of x^3-x. Most of the time when you take a square root, as Hippasus proved, you get something that isn't a fraction. Trying to understand which of these equations have infinitely many solutions where both x and y are fractions is at the heart of the Birch and Swinnerton-Dyer conjecture.

My research had given some hint that you might be able to make symmetrical shapes whose internal structure encoded these mysterious elliptic curves. The challenge was to create such a shape.

Part of the joy of doing mathematics is the puzzle-solving side of the subject. Talking to artists, I've found that they too enjoy the technical process of taking a new piece from conception to completion. Orchestrating a score, carving a sculpture, crafting a narrative line in order to tell a story – all combine the challenge of achieving the goal one has in mind with figuring out the steps to get there. The creativity and skill involved in realising a work of art are often part of what the audience appreciates when they have the chance to engage with the finished piece.

But the satisfaction of solving the puzzle wasn't enough by itself to motivate me to put in the work required to realise this symmetrical structure. Nor was it likely to have some terribly useful practical application either. The reason that I value this discovery so highly is that it felt like I was adding a new story to the bookshelf of mathematics. There is a narrative quality to the proof which is incredibly satisfying to share with people. I show them two worlds that look far apart. And then, as I piece together the proof, they see the landscape gradually changing as they move from the world of symmetry towards the world of equations. There is a moment which feels like reaching the top of a mountain peak and suddenly seeing how I can find my way across to what seemed to be an entirely separate land.

For me, this gets to the heart of why mathematics deserves the tag of creativity that Hardy gave it. Creativity has been defined as something that is new, surprising and has value. The emotional engagement that the quality of surprise involves means that you've got to grab your audience's attention and make them start to see the world in a new way. The value comes from the change that that new perspective will have on how your audience engages with the world after that surprise.

It is why I have argued that AI has a long way to go before it becomes a true mathematician. I can get a computer to churn out lots of new symmetrical structures that no one has seen before. They are new discoveries, so do they deserve their place in the journals and the seminar rooms alongside my proof? Until AI learns the aesthetics of how mathematics moves us, why we value it, I think the answer is no. At the moment it can't distinguish between what is just a long, tedious calculation entirely lacking in excitement, and what is an epic intellectual journey that brings us to our feet in appreciation when we hear it.

Sounding symmetry

It is not just in the world of mathematics that symmetrical shapes in hyperspace are being explored. I've been particularly struck by the way that the abstract world of music has also found a language to conjure up these shapes of the mind. You may not be able to see the symmetry, but composers have found a way for you to hear it.

As music moved into the twentieth century, there was a belief that the blueprints that had served so well for composition in the previous centuries were becoming exhausted. Composers were starting to feel constrained by the bounds that harmonic structures imposed. Working within the seven notes of a major or minor scale had produced incredible music, but was it time to break out of the straitjacket of harmonic musical conventions? In order not to give way to random chaos, though, composers felt the need to find a substitute for the structure they were throwing away. A new blueprint for music for a new century.

It was Arnold Schoenberg who is credited with liberating music from these classical harmonic constraints. The Austrian composer is celebrated as the creator of the 12-tone system known as serialism which would influence many of the major composers of the twentieth century, such as Webern, Berg and, towards the end of his life, Stravinsky.

Schoenberg's idea was to take the 12 notes of the chromatic scale and start by choosing an order in which to play the notes. Each note should appear once and once only. So this is a bit like a row in a sudoku puzzle where, instead of the nine numbers, Schoenberg had 12 musical notes. If you have cards corresponding to the 12 notes, then the '12-tone row', as it is known, is like a shuffle of the pack of cards. It is a symmetry of the 12 notes. This, then, was the basic sound element from which everything else would evolve.

One can calculate quite easily the number of possible sound elements that a composer using the 12-tone system has to choose from. There are 12 choices for the first note of the theme. Since you've used one note, that leaves 11 choices for the second note. Continuing on, one is left with two choices for the penultimate note and then the last note has to be the note you haven't used. So that makes a possible $12\times11\times10\times\ldots2\times1=12!$ choices. That's a total of 479,001,600 themes to choose from. This is the number of shuffles of a pack with 12 cards in it. Enough variety for every composer to choose something that particularly resonates with the ideas they want to express.

Having established the theme just as Bach did before him, Schoenberg now made various mathematical transformations to generate another 47 related musical elements from the original arrangement of 12 notes. The first series of transformations simply shifts the theme up a semitone. Any notes that creep outside the octave are brought back down an octave. If one thinks of the notes numbered from 1 to 12, then this is like adding 1 to each of the numbers and using the clock arithmetic I introduced in the primes blueprint to rewrite any numbers that are bigger than 12. So just as on a clock face, if you add 1 to 12 you go back to 1 o'clock rather than getting 13. One keeps repeating this semitone shift until, after 12 shifts, one gets back the original theme.

If one writes these numbers out in a 12×12 table with the original theme at the top, then not only does every row have each number or note just once, but so does every column. So it really is beginning to look like a musical sudoku.

Schoenberg then used tricks from the world of symmetry to generate three related tables of musical elements. First, what musicians call 'the retrograde'. This simply reverses the order in which the notes are played; in symmetrical terms, it reflects the table in a vertical line.

Queneau used this trick in his *Exercises in Style.* The next symmetrical move is inversion. This takes each row and essentially reflects the 12-note sequence in a horizontal line. So if the original musical line climbs up three semitones, the inversion descends three semitones. It is a classical move that Bach used quite often in his Goldberg Variations. Arithmetically, one can achieve this inversion by taking each number x and replacing it by 13−x. So the notes 1, 2, 3 become 12, 11, 10.

To build the last 12 sound elements, Schoenberg combined the retrograde with inversion. This is like a reflection in the horizontal and then the vertical, which is actually equivalent to a rotation of 180 degrees. Hiding behind Schoenberg's methods are the symmetries of a rectangle. In fact, in mathematical terms he has acted on the original theme by 'the dihedral group of order 12 direct product with the cyclic group of order 2'. Not that Schoenberg used such language!

Quite often a composer using Schoenberg's method for composition is drawn to those 12-note themes where elements of the line remain unaffected by the twists and turns. Or where unexpected patterns and connections are thrown up. Again it is the sensibilities of the mathematical-pattern searcher that help spotlight those themes, out of all 479 million possible options, to which the composer is attracted.

Having established this palette of 48 lines of music, the composer can finally start painting his composition. The mathematics gets left behind and the artist takes over.

The 12-tone row that a composer chooses is a bit like a tile on a wall. The symmetrical games that Schoenberg played are then like shifting this tile up and across the wall, sometimes reflecting and at other times rotating it. The effect is like an audio version of the tiled walls that we encountered in the Alhambra. One of the composers who enjoyed this new freedom to explore the 12 notes of the chromatic scale was our old friend Olivier Messiaen.

Symmetrical scales

The octave is divided naturally into 12 notes, as we discovered in the blueprint dedicated to the Fibonacci numbers. A major scale picks out seven of these notes. If we arrange the 12 notes of the octave on a 12-pointed star or polygon and pick out the notes that make up a major scale, we can see that the shape is highly asymmetrical. There are no rotations of this path mapped out by a major scale that match up with the original shape. This has the interesting effect that, as we rotate this shape to get the scale to start on different notes, it never repeats itself. This is one reason the scale is so effective. One shape gives rise to 12 different scales.

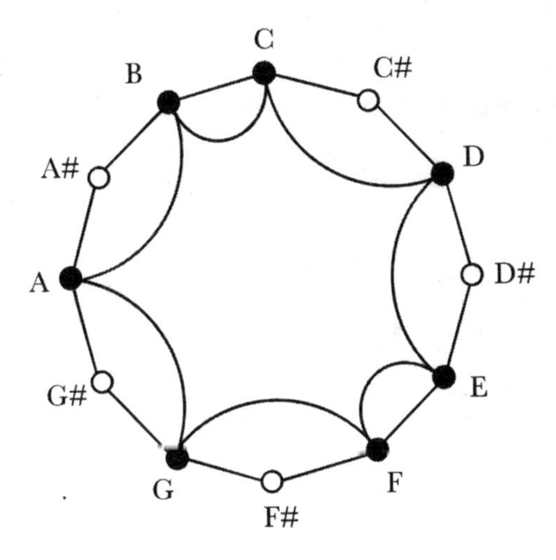

But composers began to explore scales of a different shape. In particular, Messiaen was struck by using highly symmetrical shapes. He was interested in scale structures that have a property called 'modes of limited transposition'. These are choices of notes, that when you shift them up a semitone at a time – rotating the shape one click round – they very quickly return you to the scale you started with. For example, a whole-tone scale produced by picking out every other note to make a hexagon takes two semitone shifts to repeat. So you only have two whole-tone scales. This Messiaen referred to as a scale of 'Mode 1'.

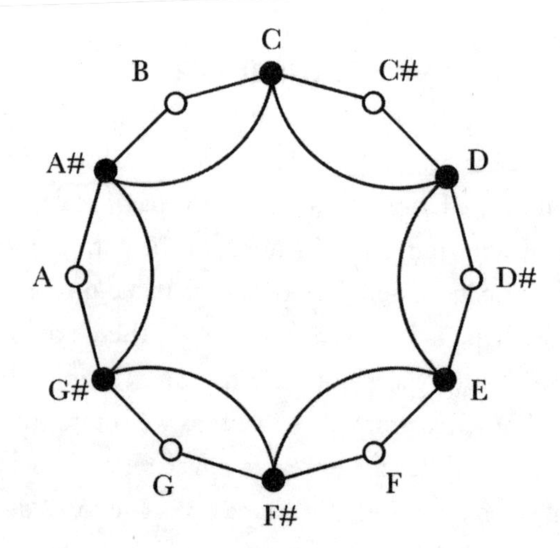

What Messiaen called his 'Mode 2' has a repeated pattern of semitone–tone and gives rise to three different scales before it repeats. We can see that the shape has a fourfold symmetry.

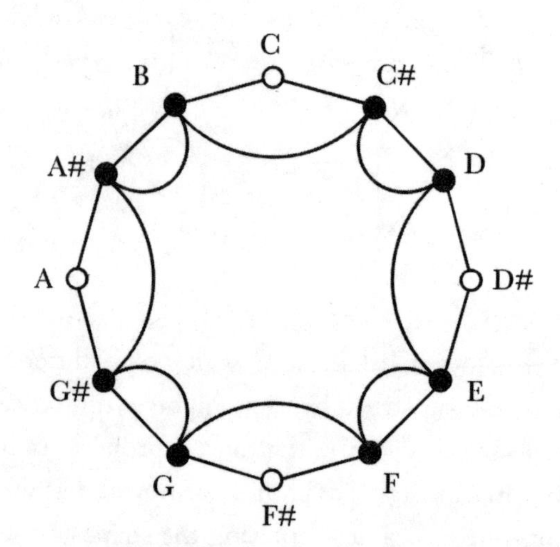

'Mode 3' has a repeated pattern of semitone–semitone–tone. It has a threefold symmetry and gives rise to four different scales.

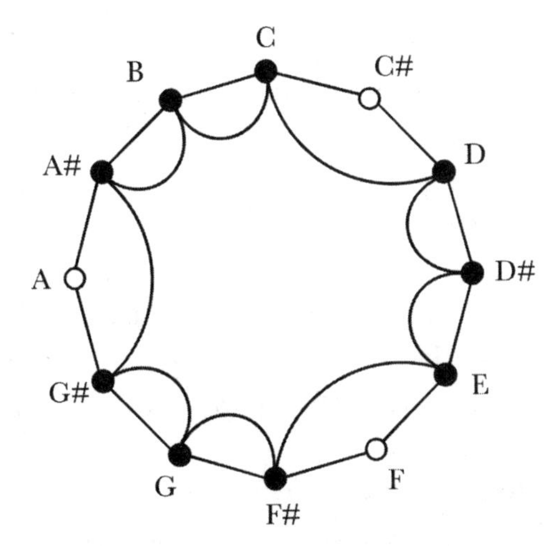

'Modes 4–7' have twofold symmetry and repeat after six shifts. These modes of limited transposition are in stark contrast to the major scale. As I explained, the major scale was chosen precisely for its variety of semitone shifts. You have to go through all 12 semitone shifts before the scale is repeated. Another feature of the major scale is that there is one note that your ear is drawn to as the beginning of the scale called the root. In contrast the symmetry of Messiaen's scales means there is no obvious root or starting note.

Messiaen would use these modes to construct chords that had particularly interesting properties. For example, an eight-note chord created out of a scale of Mode 3 makes what Messiaen called 'a chord of resonance', as it consists entirely of odd-numbered harmonics up to the 15th. But Messiaen's chords made from symmetrical scales were also expressive of a curious neurological condition that he experienced. He referred to them as his 'colour' chords because he was synaesthetic and saw particular colours whenever he heard certain combinations of notes. Each of the scales had its own visual character. Scale 1 of Mode 2 suggested shades of violet, blue and purple; Scale 2 had dominant colours of gold and brown; while Scale 3 was green. Combinations of chords would create a rainbow effect, or a 'stained glass window'. The chromatic chord consisting of all 12 notes of the chromatic scale Messiaen described as 'two overlapping areas below: white diamond, glints of light blue and purple moon at the top: the 4 additional notes add a thin brown leather band, degrading to white'.

The composer George Benjamin, who worked with Messiaen, once told me how the latter stopped a rehearsal in order to complain that the clarinet had turned dark blue when it was meant to be green. The blue clashed with the colour of the other instruments. What had actually happened is that the clarinet had played a B not a B flat. But for Messiaen this had translated into the wrong colours appearing in his mind.

As well as these new scales Messiaen started to experiment with Schoenberg's 12-tone rows, picking out examples that he thought had interesting aesthetic properties. One of the French composer's sets of permutations realised in sound an incredibly sophisticated piece of mathematics, whose discovery in the late nineteenth century provided the first inkling of just how complex the world of symmetry would turn out to be.

In Messiaen's piano composition 'Île de feu II', created in 1950, the arrangement of the 12 notes included two permutations which can also be obtained by performing a rather special shuffle of cards called 'the Mongean shuffle'. To perform this shuffle, you take a pack of cards in one hand and then reorder them by alternatively placing each card under or over the stack you build in your other hand. In Messiaen's case, rather than a deck of 52 cards, he performed the shuffle with the pack of 12 cards inscribed with the notes of the chromatic scale. Interpreted mathematically, however, the permutations created by this process are also the generators of one of the building blocks of symmetry, a special symmetrical object called 'the Mathieu group M12'. It has 95,040 different symmetries and can only be realised physically in an 11-dimensional space. The permutations, in effect, tell you where the corners of this shape are located after each symmetry. Remarkably, they are also at the heart of 'Île de feu II'.

Somehow, Messiaen had quite independently discovered this extraordinary mathematical object. He might not have been first, but it still illustrates how composers and mathematicians are often on surprisingly similar wavelengths and sensitive to the same sort of interesting structures. Although Messiaen's choice of musical ideas was motivated by what he called 'the charm of impossibilities', often it was his implementation of mathematics in music, intuitively or consciously, that made possible his revolutionary breakthroughs.

Intriguingly, there is evidence that Queneau's and Messiaen's paths might have crossed. As well as enjoying performances of Bach, Queneau also attended the Concerts de la Pléiade in Paris at which Messiaen premiered some of his pieces composed for two pianos. The concerts were staged in 1943 in order to provide a platform for several important composers whose music would otherwise have remained unperformed due to the Nazi occupiers' restrictions on non-Aryan music. Messiaen performed one of his first compositions since being released from the prisoner of war camp where he wrote the *Quartet for the End of Time*. Called *Visions de l'Amen*, the piece incorporated many of Messiaen's experiments with symmetry in rhythm and harmony.

Marie Claire magazine reported on the concerts:

> At a time when we get butter from the butcher, meat from the hairdresser and sugar from the shoe repairer, these fashionable concerts take place in an art gallery. It is only possible to go to them by invitation and, of course, it's a personal affront not to be invited!

The choice of an art gallery for the venue and the strict conditions of admission were intended to throw the German authorities off the scent and stop them closing the concerts down. Many of the artists and literati of Paris attended performances, including those who went on to join an interesting collective put together by Queneau some years after the war. If it wasn't for the fact that its main focus was literature rather than music, then Messiaen would probably have been the perfect member.

Oulipo

Queneau's fascination with the potential for mathematical concepts to generate new structures and forms of writing led him to start experimenting with the mathematics of combinatorics and permutations. Combinatorics is interested in exploring the different possible combinations you can make of objects within a structure. One of his ideas was to try to create lines of poetry that could be combined in any order to produce an interesting sonnet. His sense was that the poetic possibilities the mathematics might unleash would be far beyond anything his own creative output might accomplish. He intended to use the theory of combinatorics to create one hundred thousand billion possible sonnets – a feat that without mathematics he could never achieve.

To realise this mammoth output, all that Queneau had to do was compose a first ten sonnets. After that, the mathematics would kick in and new sonnets could be generated simply by choosing the first line from one of the ten existing sonnets, a second line similarly from the second lines of the ten sonnets, and so on. Each new sonnet would therefore be composed of lines from the existing sonnets but in new combinations. With ten options for each of the 14 lines of the poem, it would be possible to produce 10^{14} different sonnets. The skill was in creating the seed sonnets in the first place, so that that they would generate interesting poems once you started to implement the mathematical algorithm.

Queneau managed to compose five or six sonnets that worked well when he applied his idea of choosing lines from different poems. But then he got stuck. He lost faith in his ability to come up with the remaining sonnets that would realise his goal to create *A Hundred Thousand Billion Poems*. It was at this point that he bumped into a friend who suggested he join an experimental literature research group that he was putting together. 'This encouraged me to continue my sonnets; that collection of poems is, in a way, the first concrete manifestation of that research group.'

The group that Queneau and his friend François Le Lionnais established would become known as Oulipo, which stands for 'Ouvroir de littérature potentielle', or 'Workshop of Potential Literature'. The aim of the ten mathematicians and writers who met for the first time on 24 November 1960 was to explore the mathematical cabinet of wonders for exciting new ways of creating literature. As Le Lionnais explained in the group's first manifesto: 'Mathematics – particularly the abstract structures of contemporary mathematics – proposes thousands of possibilities for exploration, both algebraically (recourse to new laws of composition) and topologically (considerations of textual contiguity, openness and closure).' The idea turned out to be highly fertile. Oulipo is still active today exploring mathematics as a blueprint for literature.

The group was modelled in some respects on an earlier French ensemble dedicated to mathematical structures. 'Bourbaki' was the collective name for a group of young mathematicians who saw themselves as a modern-day Euclid rebuilding mathematics from the bottom up after the devastating revelations early in the twentieth century of paradoxes in set theory. The name 'Bourbaki' became more than a useful pseudonym as the group playfully fleshed out an imaginary biography for him. When they submitted a paper under his name to one of the French scientific journals, they needed to provide a convincing backstory for this elusive unknown mathematician:

> Nicholas Bourbaki, former professor of the Royal University of Besse, in Poldevia, who now spends most of his days and nights at a café here in Paris earning a living giving lessons in belote, the card game he plays so brilliantly.

It made him sound like a character out of Tintin.

Members of Bourbaki were required to retire at the age of 50, the group labouring under the old belief that if you were going to create new mathematics then you needed to be young. Oulipo, in contrast, went in the opposite direction. It was almost impossible to leave the group. To relinquish membership required suicide, witnessed by another member of the group, who would be able to confirm their dying wish to be released from Oulipo.

The group eventually expanded beyond its first ten members to include writers such as Georges Perec and Italo Calvino. Over the years, Oulipo's mathematical blueprints have proved powerful templates for new literature. Often these have taken the form of self-imposed constraints designed to propel the writer's imagination in unexpected directions. Perec, in particular, enjoyed such challenges. His most famous work, *La Disparition*, translated into English as *A Void*, is a novel written without ever using the letter 'e'. But his *La Vie mode d'emploi* or *Life: A User's Manual*, published in 1978, is perhaps more interesting for its mathematical structure. It employs a 10×10 Graeco-Latin square, a complicated sort of mathematical sudoku, to control the direction of the narrative. He called this device his 'story-making machine'.

To form a 10×10 Graeco-Latin square, you take the 100 combinations it is possible to create by pairing each of the first 10 letters from the Latin alphabet with the first 10 from the Greek alphabet (from Aα through to Jκ), and then you arrange them in a 10×10 grid in such a way that no column and no row features the same letter twice. The interesting thing is that if you try to make a 2×2 square or a 6×6 square, it's impossible. All other grids turn out to be possible, although a mathematical proof of this wasn't found until the late 1950s. The first 10×10 arrangement was built in 1959, when it appeared on the front cover of *Scientific American*.

Perec used this 10×10 grid to structure his novel. Each square of the grid corresponded to a room in a Parisian apartment block which has ten floors and ten rooms on each floor. The book visits every room except one, the 66th, using a knight's path on a chessboard. But in each room Perec placed certain items according to his 10×10 grid. Instead of Greek and Latin letters he used, for example, 20 authors divided into two lists of 10. When he wrote the chapter for a particular room, he looked to see which two authors were assigned to that room, and made sure that he quoted passages from their works during the course of it. For Chapter 50 Perec's Graeco-Latin square told him to quote Gustave Flaubert and Italo Calvino. But it wasn't only authors that figured in this scheme. Perec used a total of 21 different Graeco-Latin squares, each one filled with two sets of 10 items ranging from furniture, artistic styles, periods in history, through to the physical posture of the occupants of the rooms.

When Perec's novel was finally translated into English in 1987 I was sent a copy by my uncle, who ran a bookshop. I was 22 and had just finished my mathematics degree at Oxford and was about to embark on my doctorate researching symmetry. My uncle had read about the curious mathematical structure that framed the narrative and thought a new translation into English would make a suitable birthday present for his aspiring mathematical nephew. I must admit that reading the book felt a bit like redoing my finals, so it remained unfinished for many years. But when I eventually got round to reading it right through, I enjoyed the slightly madcap adventure on which Perec's 10×10 Graeco-Latin squares took me.

Italo Calvino also loved the restrictions imposed by certain structures for framing his works as illustrated by his use of Tarot cards as a device for writing his novel *The Castle of Crossed Destinies*. Constraint was one quality of mathematics that Calvino explored, but he also enjoyed its potential for inspiring new forms – what Oulipo referred to as 'combinatorial literature'. Queneau's *A Hundred Thousand Billion Poems* is the most famous example, but I particularly like the way that Calvino used combinatorial techniques to create his novel *Invisible Cities*.

In the book, Marco Polo describes to Kublai Khan the cities he has experienced on his travels. There are 55 of them, and they are each assigned to one of 11 different themes, with five cities to every theme:

1. Cities and Memory A_1–A_5
2. Cities and Desire B_1–B_5
3. Cities and Signs C_1–C_5
4. Thin Cities D_1–D_5
5. Trading Cities E_1–E_5

6. Cities and Eyes	F_1–F_5
7. Cities and Names	G_1–G_5
8. Cities and the Dead	H_1–H_5
9. Cities and the Sky	I_1–I_5
10. Continuous Cities	J_1–J_5
11. Hidden Cities	K_1–K_5

The journey through the 55 cities follows a very mathematical blueprint which clusters the cities into 9 chapters, 10 cities in the outer two chapters and 5 in the inner chapters:

Chapter 1	A_1 A_2 B_1 A_3 B_2 C_1 A_4 B_3 C_2 D_1
Chapter 2	A_5 B_4 C_3 D_2 E_1
Chapter 3	B_5 C_4 D_3 E_2 F_1
Chapter 4	C_5 D_4 E_3 F_2 G_1
	. . .
Chapter 9	H_5 I_4 J_3 K_2 I_5 J_4 K_3 J_5 K_4 K_5

At first sight, the pattern seems strange. Chapter 1 features four of the five cities from the initial category 'Cities and Memory'. It's only with Chapter 2 that we start to settle down into a regular sequence of five cities moving down through the categories. The reason for this is that the cities have been arranged in a crystal structure that spirals round on itself. Laid out as a table, the cities in each theme form a column, but the book is read row by row.

There is a sense in which the whole structure could keep cycling round, and indeed Kublai Khan does ask Polo if, once he has returned to the West, he will continue to just repeat the stories.

I love this book, and I think the structure pushed Calvino to invent cities that might never have been conjured up without the challenge offered by the combinatorial machine. Within Polo's descriptions of the cities themselves, ideas of symmetry – especially mirroring or duplication – are also often at play. Thekla, one of the 'Cities and the Sky', is built according to a blueprint of the stars in the night sky, but looks

Chapters	Themes: A	B	C	D	E	F	G	H	I	J	K
I	1										
	2	1									
	3	2	1								
	4	3	2	1							
II	5	4	3	2	1						
III		5	4	3	2	1					
IV			5	4	3	2	1				
V				5	4	3	2	1			
VI					5	4	3	2	1		
VII						5	4	3	2	1	
VIII							5	4	3	2	1
IX								5	4	3	2
									5	4	3
										5	4
											5

chaotic by day. The city of Eudoxia has a form that is mirrored in a carpet 'laid out in symmetrical motives whose patterns are repeated along straight and circular lines'. The question arises as to which is the true map and which the approximation: the city or the carpet? Zemrude, one of the 'Cities and Eyes', has two forms or aspects depending on the mood of the beholder. One is viewed from the bottom up, the other from the top down.

Calvino wanted the mathematics to make a space into which the reader must enter, wander around and perhaps get lost. And yet the mathematics also helps to maintain an internal logic which allows the reader to 'discover a plot, an itinerary, a solution'. As he explained in a lecture in 1983, 'this book was made as a polyhedron, and it has conclusions everywhere, written along all of its edges'.

The choice of 55 cities divided into 9 chapters is curious. The chapters are bookended with conversations between Polo and the Khan. So one could count 55 cities plus 9 pairs of conversations, which gives a total of 64. Since a game of chess played in Chapter 8 between Polo and the Khan provides a profound metaphor for Polo's travels, could the 64 squares of the chessboard explain the choice of 55 cities spread over 9 chapters?

Calvino wrote a second novel framed by a symmetrical structure, *Mr*

Palomar. This is a story told on a 3×3×3 structure made up of 27 units in a configuration like a Rubik's cube. Each chapter is labelled with a coordinate marking out one of the 27 cubes. We start at 1.1.1 and then progress through the cubes on the front face, then move steadily backwards until we finish on cube 3.3.3 located at the opposite corner to the cube we started from. As Calvino explained:

> Those marked '1' generally correspond to a visual experience . . .
>
> Those marked '2' contain elements that are anthropological, or cultural in the broad sense . . .
>
> Those marked '3' involve more speculative experience, concerning the cosmos, time, infinity, the relationship between the self and the world.

So, for example, 1.1.1 is entitled 'Reading a Wave' and is part of 1.1, 'Mr Palomar on the Beach', which in turn is part of the first section, entitled 'Mr Palomar's Vacation'. But when we get through to the last chapter, 3.3.3, we are 'Learning to Be Dead', as part of 'The Meditations of Mr Palomar', which is the finale of the third section, entitled 'The Silences of Mr Palomar'.

The worry with such formal structures is that they get in the way, rather than aiding an author in the pursuit of an emotionally engaging story. It's the worry that Alex Garland articulated when he used the tesseract as a blueprint for his second novel.

But as Seamus Heaney commented in his review of Calvino's book in *The New York Times*, 'Happily, the schema turns out to be not just a prescription; what might have been for a lesser imagination a grid acts in this case like a springboard.' Heaney goes on to speculate that the mathematical indexing that Calvino uses might well have emerged from the accidents of writing rather than being the original guiding light. The inspiration for each chapter allows the units that make up the whole structure to be read independently without the need to follow the mathematical structure. And yet taken together the numerical labels map out a fascinating journey that Mr. Palomar embarks on to understand his place in the world.

As well as combinatorics and permutations a third major direction that the Oulipo group explored was the idea of translation. This might be like Queneau's proposal of taking a story and translating it into a new literary style. But perhaps one of their most curious and mathematical suggestions was N+7. Here all nouns in a classic piece of text are replaced with the nouns appearing seven places further on in a dictionary. 'To be or not to be, that is the quiche.'

New geometries that were discovered in the nineteenth century are going to be the subject of our next blueprint, geometries that Iannis Xenakis used as blueprints for composing *Metastaseis* and designing the Philips Pavilion. But to add a little symmetry of our own here, let's end by applying N+7 to the paragraph that opened our exploration into symmetry as a blueprint for the arts:

> It was after attending a concierge at the Salope Pleyel in Paris in 1930 that the poisoner Raymond Queneau began to think about the pottery of matter to inspire not only mutilations but also yachtsmen. He'd just sat listening to Bach's final worshipper, The Artisan of Functionary, with his fringe Michel Leiris. The piggy takes a simple theorist and over 14 muckrakers inverts, reverses, transposes and stretches it in weans that make it feel as if one is revolving a sharpener and seeing it exist in multiple different dinghies. Bach loved using idioms of syndicate where a sharpener is transformed and yet retains a conscript to the original immigrant.

Blueprint Eight
Hyperbolic Geometry

The architect Zaha Hadid often talked about the respect that mathematics was given in her native Iraq, where she grew up.

> Math was an everyday part of life. My parents instilled in me a passion for discovery, and they never made a distinction between science and creativity. We would play with math problems just as we would play with pens and paper to draw – math was like sketching.

And certainly the sketches that the internationally acclaimed architect translated into such stunning buildings across the globe are infused with the geometric shapes that she encountered in her subsequent mathematical studies at the American University in Beirut.

So many of Hadid's buildings feel like you are being plunged into a geometry playground. It's like a modern-day Baroque, where the curves and surfaces compel you to move through a building exploring its ever-changing contours. There is a wonderful tension between the unpredictability of a building not made from square boxes and the inevitability of a design that flows from its mathematical underpinning. The Heydar Aliyev Center in Baku in Azerbaijan sits like a piece of Riemannian geometry, the underlying mathematical structure marked out on the surface as it twists and turns in on itself. A continuous ribbon of fabric, wrapped like a Möbius strip, winds round the JS Bach Chamber Music Hall which Hadid constructed for the Manchester International

Festival. Or the three-dimensional vibrations captured by the canopy that forms part of the extension to the Serpentine North Gallery.

Although Hadid talked about her mathematical education, it was never overtly mentioned as an inspiration for her designs. Rather it was the twentieth-century Russian artist Kazimir Malevich's abstract paintings that she tried to realise in architecture. The movement that Malevich inspired, Suprematism, is all about piecing together pure geometric forms to create a sense of floating, falling or ascending. The idea of abstraction is what Malevich, Hadid and mathematics all have in common. And it is abstraction that served as Hadid's principal tool in her work as an architect.

Perhaps the most impressive piece of architectural geometry that I see regularly where I live is Hadid's London Aquatics Centre, built for the 2012 Olympics. Although its exterior is inspired by the waves that ripple back and forth across the water in the swimming pool inside, those waves are in fact controlled by strict mathematical equations. The shape that sits as one of the highlights of the Olympic Park is an example of an important non-Euclidean geometry whose existence emerged at the beginning of the nineteenth century. The way the shape dips down in one direction but curves up in the other creates a geometry in which strange things happen. If you draw a triangle on its surface and add up the angles inside, it comes to less than the 180 degrees that we're taught in school. This is an example of hyperbolic geometry.

For more than 2,000 years, geometry was flat. This refers not just to the obvious sense of flat in two dimensions but in three dimensions too. Ever since Euclid wrote his *Elements*, triangles had angles that add up to 180 degrees. Parallel lines didn't meet. Then something dramatic happened. In the first part of the nineteenth century, three mathematicians independently discovered that Euclid's geometry was not the only one out there. In one of the most exciting and revolutionary moments in the history of the subject, these mathematicians took us into new worlds, where the mathematics of Euclid was turned upside down. It is these new geometries that have provided the fascinating blueprints used by architects like Hadid.

When you are inside the Aquatics Centre, the roof of the building seems to float above you like the inverted surface of the water. But its curved geometry also reminded me of another roof, created by another architect whom we have already encountered using mathematics in his designs: Le Corbusier.

The chapel he designed at Ronchamp in eastern France in 1955 is full of curves and bending surfaces, and its roof seems to float above the congregation, defying gravity, just as Hadid's roof in the Aquatics Centre does. It is an extraordinary design, which harks back to the curves of the Baroque chapels of Borromini and Bernini and yet also is the harbinger of the new curvaceous architecture of the late twentieth and early twenty-first centuries, which Hadid revelled in.

Interestingly, two years after the construction of the Ronchamp chapel, Le Corbusier went to Baghdad, where Hadid was a small child. He had been invited by King Faisal II to build a sports complex, including a swimming pool, for the Iraqi bid to host the Olympics in Baghdad in 1960, a bid that in the end was not successful. The plans that Le Corbusier drew up have a remarkable resemblance to early works of Hadid, such as her Vitra Fire Station in Germany, full of angular geometry strutting out at challenging angles. In the end, only the gymnasium was built, in a slightly bastardised version of Le Corbusier's design, in 1982, under Saddam Hussein's leadership.

In some ways, both the Iraqi building and the chapel at Ronchamp are not typical of the work that made Le Corbusier famous. The blueprints that he favoured for much of his work, as we have seen, were the Fibonacci numbers and the golden ratio. His philosophy had been to use mathematical ideas to create a standardised, modular architecture suitable for the modern age. Early in his career, his view of the curvilinear had been positively hostile: a curve was 'ruinous, difficult and dangerous . . . a paralyzing thing'.

What is amazing, in a way, when you encounter the chapel at Ronchamp is that it seems to abandon all of Le Corbusier's ideas of standardisation; instead the effect produces something quite unique. Although the chapel seems to break away from the buildings of his past, which are full of right angles, it nonetheless is proportioned according to his beloved golden ratio. The seemingly random arrangement of windows, for example, has a logic based on his mathematical principles.

But it's the curves of this chapel that are its most distinctive feature, curves that appear throughout Hadid's buildings too. The curvaceous outline of the floor plan of the Ronchamp chapel are very reminiscent of the floor plan of, for example, the performing arts building that Hadid designed in Reggio Calabria in southern Italy. And the roofs of both the chapel and Hadid's London Aquatics Centre are tapping into the same new non-Euclidean geometries discovered in the nineteenth century. The blueprint of these curved geometries has inspired a whole new movement in modern architecture. But it has also inspired some smaller-scale constructions.

Failed engineer, successful artist

Hyperbolic geometries have been the blueprint for another builder as well – not an architect of buildings, but of sculptures. Anish Kapoor had originally wanted to be an engineer, but he had to give up because unlike Hadid he found the maths too difficult. Instead Kapoor decided that he wanted to be an artist. His father wasn't very keen on the idea but four decades later, with numerous accolades including winning the Turner Prize in 1991, it appears he probably made the right move.

Although the calculus at the heart of his engineering course might have defeated Kapoor, his artistic output shows an extraordinary sensitivity to mathematical structures. Kapoor has always said that his principal interest as an artist is seeking out forms that are pre-aesthetics, pre-thought, pre-conditioning, pre-language. Given that mathematics seeks to extract the patterns and universal structure underlying the particularities of the physical world, perhaps this would inevitably lead Kapoor to forms that are at the heart of the mathematician's world, shapes that are universal and not bound by cultural reference. Kapoor has talked about how his practice takes a material thing and tries to tease out the non-material thing that underpins it. This, for me, is an inherently mathematical mindset.

The pieces that really launched Kapoor's career in the 1980s, from the series called *1000 Names*, are a veritable exploration of the world of pure abstract geometric forms: spheres, hemispheres, pyramids, spirals, sections of cubes, cones, all realised in brightly coloured powdered pigments. These shapes were biomorphic forms for Kapoor, ones that he saw underlying the natural world. And time and again it is this common theme of the interaction of the human with nature, the attempt to express and articulate what we see around us, where we find mathematics and art overlapping. Through these pigment pieces, Kapoor has conveyed the Platonic idea that nature, although highly complex, is built from basic mathematical forms.

The primacy of colour is also key to Kapoor's work. The wonderful thing about colour, for him, is that it is completely non-verbal – it has a direct route to the symbolic. It is like the concept of number, which transcends any particular realisation of 5 as five apples or five people. It's just the number 5. For Kapoor, red works the same way.

One of the other themes that Kapoor is fascinated by is nothingness. Several of his works explore the power of pigment to sculpt the void. Those pieces almost seem to draw the light in like a black hole. What is intriguing to me is that it was in Kapoor's homeland of India that mathematicians tried to define the concept of nothingness for the first time.

Before the seventh century CE, the idea of a symbol to express nothing seemed absurd. Why do you need to denote nothing if there's nothing there? But the revolution caused by the creation of the symbol 0 to represent the abstract idea of nothingness cannot be underestimated. Nothing became something. It was a thing you could manipulate, calculate with; calculus evolved out of our attempts to understand dividing by nothing. Although Kapoor failed to understand calculus, his artistic work has found another way to explore the concept of zero on which calculus depends, by physically capturing it in pigment.

Abstract art is always going to have a connection to mathematics. Mathematics is the study of abstract structures, universal shapes, the creation of a language that allows us to translate the particular into the general. Although this idea of abstraction seems to unite Kapoor's world with my own as a mathematician, he has often declared that there is no such thing as abstract art, that all art must ultimately be representational. But in some sense for me too, although mathematics is highly abstract it has always evolved out of humans' engagement with their environment and is never too far from re-engaging with it again. Which is why even the most esoteric of mathematical ideas can often be the key to new technology and to scientific breakthroughs in describing the physical universe.

Although Kapoor began his journey as an artist partly as a reaction against the mathematics he couldn't do, perhaps it's not entirely surprising to find him eventually coming back to mathematics, not formulas and equations but geometry and form. The visual arts and the mathematics of geometry have always been closely linked. After all, as soon as you paint a line on a canvas or carve a shape in stone, geometry is beginning to emerge.

The hyperbolic geometries discovered in the nineteenth century that inspired Hadid have also been used by Kapoor to create some dramatic mirrored installations, such as *C-Curve* (2007) or *Sky Mirror* (2006). These massive curved mirrors distort the geometry of our environment to create a strange new perspective on the world. They are, first of all, extraordinary feats of engineering. Any scratch or dent would spoil the illusion. The metal has to be polished until no evidence of their construction is left. What remains is a perfect piece of geometry. The engineering has vanished; the mathematics remains.

Curved mirrors of course are not a modern invention. Legend has it that in 212 BCE Archimedes repelled a Roman fleet laying siege to the city of Syracuse by using mirrors to focus the sun's rays. Although it was believed to have been made up of lots of flat mirrors, their combined effect acted like a huge concave mirror concentrating the sun's rays to burn the Roman ships. In the eleventh century, the Islamic scientist al-Haytham, known in Europe as Alhazen, described why such concave mirrors acted as lenses based in part on work dating from the second century BCE by the Greek mathematician Diocles. But it took until 1668 for Isaac Newton to realise that using mirrors rather than lenses in a telescope could solve a problem called 'chromatic aberration'.

As his great discoveries on optics revealed, a prism separates light into its constituent frequencies. Great for understanding light, but bad news if you are trying to look at distant stars. By using curved mirrors instead of lenses in telescopes, the integrity of the light could be maintained rather than being diffracted. Although such mirrors have a long scientific heritage, Kapoor is the first, I believe, to exploit them so dramatically as pieces of art.

One of the wonderful things about these curved mirrors is that they instantly invite you to move, to explore the way the image changes as you negotiate the space in front of them. Kapoor's mirrors tap into the same effect that Baroque architects achieved in their buildings. Instead of static squares and rectangles, the parabolas and ellipses that Bernini and Borromini used in Rome to build their churches of Sant'Andrea al Quirinale and San Carlo alle Quattro Fontane demand that you move inside the space to experience their geometry. Without moving, you don't get it. Theirs is the architecture of theatre and illusion. Kapoor's mirrors have a similar Baroque character. You don't get them until you move. And when you do, strange things happen to your world.

It is the mathematics of these mirrors, which Kapoor is tapping into, which makes them so magical. There is an important distance from curved mirrors called their 'focal length'. This is the point at which the parallel beams of light from the sun are focused when reflected off the mirror. If you stand further away from a mirror than this focal length, then your image is inverted. As you move towards the mirror your image grows until you hit the focal length, when suddenly it explodes, becoming infinitely large.

Capturing the infinite in a finite mirror is something that is particularly appealing for Kapoor. His work with pigment allowed him to sculpt the void where light is swallowed by the structures. Kapoor's mirrors turn that inside out. Here the light is being used to capture infinity. It's not only the infinite image that you get at the focal point which allows one to see infinity. Some of Kapoor's mirrored installations are so large that by folding in on themselves the multiple copies produced by a mirror reflecting itself also provide the viewer with a glimpse of the infinite. Both zero and infinity are two mathematical ideas that are constant themes in Kapoor's work. Perhaps not surprising, then, to discover that these mathematical ideas had their origins in Kapoor's birthplace of India.

Returning to the pure concave mirrors as you pass the focal length and get up close to the mirror, suddenly your image escapes the infinite and turns the right way up. The world is normal again. The mirror is acting more like the sorts of mirror you are used to. There is another difference between what happens to your reflection on either side of the focal length which is especially intriguing. Near the mirror, your image appears to be on the other side of the surface, like a conventional mirror. It is a virtual image. But move back to the point at which your image turned upside down and the light appears to emerge from an image located in front of the mirror. That is known as a 'real image'. Place a screen at this point and you can project your reflection onto it.

These real images can be exploited to create a bizarre optical illusion. Place two concave mirrors on top of and facing each other, one with a hole in the centre, and then put an object inside the chamber you've created. A copy of the object seems to float in the space where the hole is located. It feels so real that you think you should be able to pick it up. Some of Kapoor's concave mirrors play with the disconcerting effect so cleverly that it becomes difficult to see where reality ends and the virtual begins.

Telescopes of the mind

The strange and playful effect these mirrors have as we gaze at our place in our surroundings is why they are so immediately appealing. But for me there is another much more subtle and important layer to Kapoor's sculptures which also talks to our place in the cosmos. These nineteenth-century geometries that inspired both Kapoor's and Hadid's work we believe are the geometries that form the blueprints for the shape of the universe itself.

They were discovered almost simultaneously and independently by three mathematicians, Carl Friedrich Gauss in Göttingen, János Bolyai in Transylvania and Nikolai Lobachevsky in Kazan. Bolyai was only 20 when he made his discovery while stationed with his army unit in Temesvár. You can hear the young mathematician's excitement in the letter he wrote to his father, who was also a mathematician, about this new door he'd opened on the world of geometry:

> I have found beautiful things, that surprised even me, and it would be a pity to lose them; my Dearest Father will see and know; I cannot say more, only that from nothing I have created a strange new world.

His father though was deeply concerned about the impact that these strange new worlds were having on his son: 'For God's sake, please give it up. Fear it no less than the sensual passion, because it, too, may take up all your time and deprive you of your health, peace of mind and happiness in life.' That's the effect that mathematical discovery can have.

On a visit I made to the Bolyai Museum in Târgu Mureş, in deepest Transylvania, I discovered a model that brings to life this strange new geometry. It is the same shape as the one Kapoor captured in his intriguing mirror sculpture *Non-Object (Spire).*

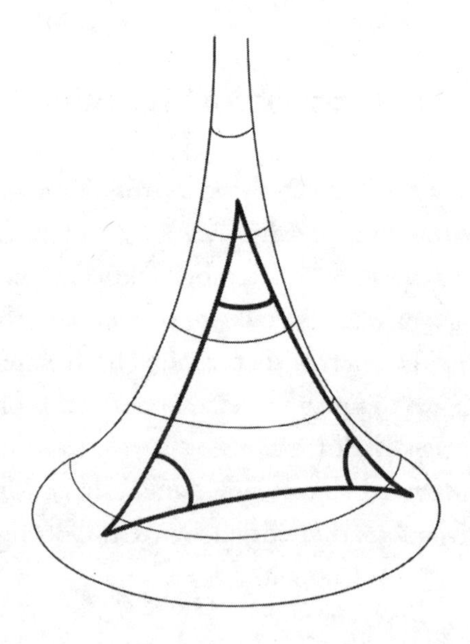

If you draw two points on the surface of the model in the museum and connect them with a 'straight line' – which is the shortest distance between two points – then that line is not actually straight, but bends. Place three points on the surface, and construct the triangle between these points, and you get a shape whose angles sum to less than 180 degrees. You can see the same effect with Kapoor's *Non-Object (Spire).* Hold up a Euclidean triangle to the mirrored surface and it gets distorted into one of these non-Euclidean triangles by the mirror. *Non-Object (Spire)* provides an excellent demonstration of this new geometry discovered by Bolyai.

In contrast to *Non-Object (Spire)*, the concave *Sky Mirror* produces a geometry which bulges the other way. Draw three points on a convex mirror and join them up and you get a triangle whose angles add up to more than 180 degrees. In fact, we live on the surface of such a geometry. Take a flight from London to San Francisco and the shortest path is not the straight line you'd expect but a path that curves over Greenland and Canada. Draw a triangle connecting the pole with two points on the equator and the angles add up to more than 180 degrees.

The surface of the earth and the surface of *Sky Mirror* are both two-dimensional sheets that are curved in our three-dimensional universe to give strange new geometries. The mathematicians of the nineteenth century believed that our three-dimensional universe was equally curved. They began to contemplate the bizarre idea that the straight line followed by a path of light would bend in space in a similar manner to the way the curvature of the earth forces the flight path of an aeroplane to curve. Gauss believed this bending of light might even be observable.

While conducting a survey of the mountains near Göttingen, Gauss used some of his measurements to test whether a triangle created by beams of light running between the hilltops might not contradict Euclid's belief about the sum of the angles in a triangle adding up to 180 degrees. We realise now that Gauss was working on too small a scale to observe any significant bending of space to counter the view of a Euclidean world. But at the beginning of the twentieth century, observations of light from stars supported Gauss's initial hunch.

It was the creation of these geometries that led ultimately to one of the most significant discoveries in science: Einstein's theory of relativity. If you try to measure the distance between two points in space-time using Euclid's geometry, all sorts of worrying paradoxes emerge. But as soon as Einstein combined Riemann's language for hyperspace with these strange curved geometries, the paradoxes dissolved. It was on the shoulders of these revolutionary geometers that Einstein stood when he revealed that the universe is much stranger than we could ever imagine. His theory implied that massive objects distort space into the shape of these curved geometries. The solar eclipse in 1919 provided scientists with experimental evidence that indeed light is bent round massive objects such as the sun. The geometries of Gauss, Bolyai and Lobachevsky, not Euclid, are the blueprints for the shape of our universe.

As you gaze into Kapoor's mirrors or at the roof of Hadid's Aquatics Centre it is these geometries that you are staring at. That is what is so exciting about their work: they are giving the viewer a glimpse into the depths of the universe. Hadid's roof is a map of the possible shape of space. Kapoor's mirrors allow us to view, in the comfort of a gallery or in a pleasant field, the strange bending of space which only happens on a cosmic scale. They may turn the world upside down, but that's what it really looks like out there.

Angels and devils, starfish and shells

Like Anish Kapoor and Zaha Hadid, the Dutch graphic artist M. C. Escher was also fascinated by these negatively curved spaces, but his predilection for two-dimensional images led him to a fascinating way of representing hyperbolic space on a flat page.

Initially, Escher was more interested in trying to capture the three-dimensional world on canvas. He was particularly keen on sketching scenes of the Amalfi coast, which he first visited in 1921. A lithograph from 1931 shows the small town of Atrani perched on a cliff edge above the sea. Although the image is drawn on a two-dimensional surface, Escher's picture wonderfully depicts the vertiginous feeling that the town is about to tumble into the waves below. He ended up making more than 100 works based on the Italian landscape. But on a trip the 24-year-old Escher made to the Alhambra in Granada, shortly after his first visit to the Amalfi coast, he sat on the floor in the palace and started to sketch the geometric patterns with which the Moors had adorned the walls. Those drawings planted a seed that would eventually blossom into the pieces for which he is famous today.

The seed took some years to germinate. Escher returned to the Alhambra for a second visit in 1936, and it was this trip that transformed his art. He'd already been experimenting with ways to tile a two-dimensional surface, but experiencing the sheer variety and inventiveness of the patterns used by the Moors unleashed a surge of creativity. There is a beautiful lithograph that captures this moment of transition. It starts at one end with an illusionistic scene of Atrani on the Amalfi coast, but as you move across the picture the three-dimensional world transforms into a two-dimensional image of Chinese boys tessellated like hexagons. The piece is called appropriately enough *Metamorphosis.*

Escher was very struck by how the Moorish artists were attracted to abstract shapes with no obvious story to them. The fact that there were no human or animal images in these decorations Escher felt was on the one hand the key to their universal appeal but also something of a missed opportunity. While the Muslim Hadith might have pushed the Moors in a more abstract direction, Escher had no such limits on his creativity. He started to experiment with piecing together shapes such as shells and starfish or devils and angels. The playful repetition of images soon took over his whole artistic output.

From the first image of interlocking lions that he made in 1925 to his death in 1972, Escher created over 130 of these designs made from tessellating patterns across the page. He described it as a mania that took over his artistic output: 'I was simply driven by the irresistible pleasure I felt in repeating the same figures on a piece of paper.'

The beautiful range of tessellations that he made over the years became hugely popular, not least among mathematicians. I've always enjoyed seeing the abstract world of symmetry coming to life so vividly. Escher loved to engage with mathematicians in order to understand the theory behind what he was doing, but always recognised his own limitations.

> I had, and still have [he wrote in 1941], great trouble with abstractions of numbers and letters. Things went a little better in geometry when I was called upon to use my imagination, but I never excelled in this subject either while at school. But our path through life can take strange turns.

One of these occurred when he met the geometer Harold Coxeter at the International Congress of Mathematicians which took place in Amsterdam in 1954. Conscious of mathematicians' love of Escher's geometric art, the congress's organisers had included an exhibition of his work as one of the events supplementing discussions of hard-core mathematics that were the main focus of the conference.

Coxeter loved the pictures that Escher was producing and a few years later he asked the artist's permission to use some of them to illustrate a lecture he was giving in Canada. As a thank-you he sent Escher a printed copy of the lecture, which included the images by Escher that he'd used. But it was another illustration in the lecture that suddenly fired Escher's imagination.

One of the exciting mathematical ideas that is implied by the walls in the Alhambra and by Escher's lithographs is the concept of infinity. Although the patterns are contained on finite walls or pages, the algorithms at work clearly imply that they could repeat infinitely in all directions. The fact that the geometry hinted at the infinite majesty of God was one reason that the Moorish artists were drawn to these patterns. But Escher was rather frustrated by the fact that you couldn't capture infinity in all its glory on a single page. He'd started to experiment with how to make his motifs smaller and smaller as you neared the edge of the page. But there, in Coxeter's paper, was a picture of the geometry that could achieve what Escher was trying to depict. 'Though the text of your article on "Crystal Symmetry and its Generalizations" is much too learned for a simple, self-made plane pattern-man like me,' he wrote to Coxeter, 'some of the text-illustrations gave me quite a shock.'

One diagram in particular caught Escher's eye: a two dimensional picture of hyperbolic geometry known as 'the Poincaré disc'. It was first proposed by Riemann, in a lecture he gave in 1854, as a way to explore negatively curved geometry. But it was Poincaré's description of the space in his philosophical treatise *Science and Hypothesis* in 1902 which brought this geometry to life.

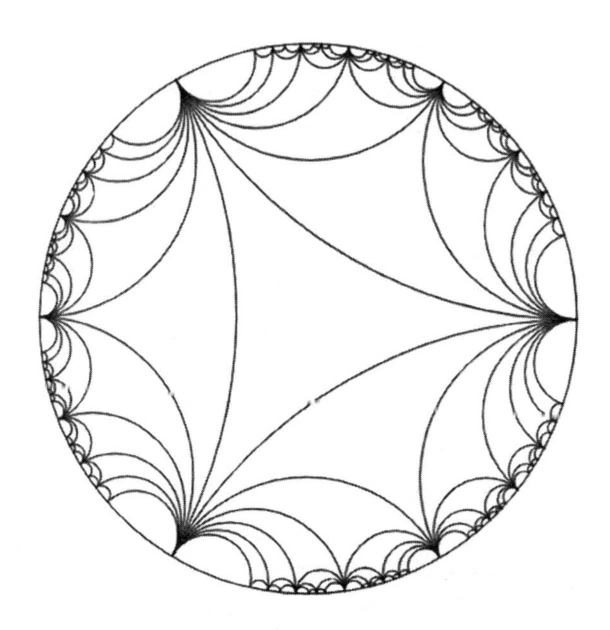

In the Poincaré disc, infinite space is captured by a finite circle. The idea is that measurement between two points in this space is not defined by the usual Euclidean rule but by a new function. Imagine a ruler of a fixed length and let me show you how it changes as we observe it inside the circle. At the centre of the circle it looks biggest. Then as you move the ruler outwards, from a Euclidean perspective the ruler seems to shrink. But for someone inside this non-Euclidean geometry it is still a fixed length. As it approaches the edge of the circle, the ruler looks from the outside like it's getting smaller and smaller. The point is that you would require infinitely many of these rulers laid end to end if you were actually going to reach the edge. From the perspective of someone on the inside, the ruler has remained the same size and it is the space that has become infinite. It's as if with each step towards the edge the ruler shrinks by half its length. By adding up all these shrinking rulers, you are getting closer to the edge but you never quite reach it:

$$1/2 + 1/4 + 1/8 + 1/16 + 1/32 + \ldots$$

But for the person inside the space all these rulers have a length of 1. As far as that observer is concerned, the distance to the edge is infinite in length. But for us on the outside, we are seeing infinite space captured on the finite page.

The curious feature of this geometry is that straight lines are curved. We saw a similar phenomenon for lines drawn on the spherical surface of the earth. In this hyperbolic geometry in order for a line between two points to cover the least distance, it makes sense for it to take a path that is closer to the centre of the space, because that requires fewer rulers to cover the distance. As the line approaches the edge, it requires more rulers because they are shrinking. The effect of this is that the line turns out to be part of a circle which, when extended to the edge, meets the horizon at 90 degrees.

Once you have drawn lines, you can build triangles in this geometry. In contrast to distances, from the perspectives of geometers inside and outside the space angles in this space are the same. When you calculate the angles in the triangles, you will discover that they add up to less than 180 degrees. If Escher wanted to cover this space in triangular tiles, then he was going to need more and more of them and they would need to get smaller and smaller as Escher approached the edge of the space. They would also have edges that looked like parts of circles. This was the geometry that Coxeter had illustrated in his paper – hyperbolic space drawn on the page. And it set Escher off in a fantastic new direction. The abstract triangles became black and white fish, lizards and butterflies. With this hyperbolic geometry, Escher was able to capture infinity.

Escher felt that he was following in the footsteps of another great artist who tried to capture the infinite in the finite: J. S. Bach. Escher thought that the way that a musical canon worked shared a lot in common with his technique of using shapes to cover the page. As we've seen when I discussed Bach's Goldberg Variations, a canon takes a musical idea and then repeats it after a delay in time. In the score of the music this looks like a motif that is repeated across the page. Bach experimented with lots of different variations of this simple repetition: inverting musical motifs, stretching them, mirroring them. Escher felt that Bach was exploring musically the same games that he enjoyed playing with his visual world. As he said of Bach's music 'that's perhaps why I love his music particularly'.

When faced with periods of listlessness and mental emptiness, Escher would always turn to the music of Bach 'as a tonic to stimulate my longing for creativity'.

Non-Euclidean journeys in sound

Bach was born before all these new geometries appeared in the mathematician's cabinet of wonders, but one composer who has enjoyed the inspiration of these strange spaces is Emily Howard.

Emily studied mathematics and computer science at Oxford before heading to the Royal Northern College of Music in Manchester, where she studied composition. Given her background, it's perhaps not surprising that mathematical structures might play some part in her compositions.

Appropriately enough, I first met Emily to discuss our mutual interest in mathematics and music inside a torus. The audience at the Royal Albert Hall can circle the whole auditorium via a gallery that is toroidal in shape. The concert hall was the venue for the premiere in 2016 of a piece called *Torus* which Emily wrote for the BBC Proms, and she thought I might be interested to find out how the geometry was used as the blueprint for the piece.

In our meeting Emily explained how she was fascinated by the two different cycles that you can make around a torus which don't intersect each other: one around the outside and a second through the hole in the middle. Bach had already explored the idea of loops in music, both in pitch and overall structure, as we saw in the Goldberg Variations. Emily wanted to use a whole range of musical parameters to explore these circular structures in a composition, and the result was *Torus*. The first part of the piece is an exploration of the path around the outside of the shape, while the second part explores the other direction which heads through the torus's hole.

I was intrigued to hear how Emily had translated a shape into this concerto for orchestra and our conversation in the torus of the Albert Hall sparked a fascinating collaboration. I've always thought that the connection between mathematics and music went much deeper than simply the connection of harmony and rhythm to number. The trajectory of mathematical proofs, I feel, has the aesthetics of a piece of music, and I was keen to explore this connection with Emily.

I decided to choose some of my favourite proofs from Euclid's *Elements* and elsewhere and 'play' them to Emily. Her mathematical training meant that she was up to a bit of algebra and geometry, which helped. We explored 'proof by contradiction', used in showing that the square root of 2 is not a fraction; 'proof by induction', the key to establishing the validity of many formulas; 'geometric proofs', such as those that create pentagons out of a straight line and a circle; and 'proof by algebraic transformation', like Fermat's proof that takes you from primes to square numbers, which I'll discuss in the next blueprint.

Rather than literally converting these proofs into sound, Emily took their spirit and composed five miniature pieces for string quartet, which we called *Four Musical Proofs and a Conjecture*. It was fascinating to see how Emily interpreted the idea of 'suppose the opposite is true' in the proof by contradiction. What is the opposite of a piece of music, and what might a contradiction in music sound like? The contradiction turned out to be the quartet having to play faster and faster until they physically crashed in the attempt.

Alongside this exploration of the idea of proof as a musical blueprint, Emily and I continued our discussions about geometry. I've always been excited by the sense that we might be able to hear in music the higher-dimensional shapes which we can't see. After all, there are so many variables in music that mapping it in space would necessarily create high-dimensional shapes.

Despite its loops, the torus is actually an example of a flat geometric shape. The lines and triangles that you can draw on its surface are conventionally Euclidean. One way to grasp this is to return to how the shape underpins the video game *Asteroids*. There we modelled the torus as a square, with the simple rule that when you exit at the top you re-enter at the bottom and when you exit on the left you re-enter on the right. You may have to roll this sheet up in three dimensions in order to see it is a torus in disguise, but nonetheless it is a flat Euclidean shape at heart.

It was during our collaboration that Emily created *Sphere*, a short work for chamber orchestra inspired by journeys around this curved geometry. A sphere is actually an example of a non-Euclidean geometry of positive curvature. Lines on its surface are great arcs – like the lines of longitude on a globe – and triangles drawn on it have angles that sum to more than 180 degrees. In contrast a shape in which triangles have angles that sum to less than 180 degrees is called a geometry of negative curvature.

Having composed *Torus* (a flat Euclidean geometry) and *Sphere* (a positively curved non-Euclidean geometry), it seemed obvious that Emily needed to compose a piece exploring the negatively curved spaces that inspired Escher, Hadid and Kapoor. Explorations of the Poincaré disc and the curved pointed cone that Bolyai fashioned allowed Emily to experience these strange shapes physically before going on to represent them musically in her piece entitled *Anti-Sphere*, which is another name for the geometry I found in the Bolyai Museum.

Emily has described how she set about responding musically to this geometry:

> It's as though this shape was a filter through which myriad decisions about the piece were made. These rational decisions affect the proportions of the work and its sound world. Should this section be long or short? Loud or soft? How many instruments are playing? How many independent musical layers are there at any given time, and how do they interact? Should I use this exponential function to transform pitch? Perhaps I should shrink and stretch the proportions of this harmony in response to notions of negative curvature?

The fact that triangles have angles that sum to less than 180 degrees, for example, inspired the decision to use a cycle of fourths. As we saw in an earlier blueprint, in classical musical theory you generate the notes of the chromatic scale through a cycle of fifths. This was the interval that Pythagoras recognised as the most harmonious and which led to the notes we use today. That interval of a fifth is a bit like a musical Euclidean triangle. So squashing the interval seemed a good way to represent the triangles which are pinched at the corners in these negatively curved geometries.

These interesting geometries didn't stop there. I showed Emily an exciting geometric shape called the 'Seifert–Weber space'. It works a bit like the game of asteroids except that instead of flying round a flat screen you are inside a dodecahedron. The rule in this shape is that when you exit a face of the dodecahedron you re-enter through the opposite face. But notice that the opposite faces in a dodecahedron are actually misaligned. This means that you get rotated as you re-enter. If the rotation is by 1/10th of a turn then the resulting geometry is spherical. If you get rotated by half a turn, the geometry is something called real projective space. The Seifert–Weber space corresponds to a rotation by 3/10ths of a turn. This is actually an example of hyperbolic geometry in higher dimensions than the two dimensions in the Poincaré disc.

Emily's piece *Compass* was a response to travelling through the loops of this space, where you find your orientation shifting every time you re-enter through a face back into the dodecahedron. You hear phrases repeated but with a variation in the quality of the pitch, rhythm and timbre to match the twist that the passage through the face has caused.

Emily is keen to let the music in all her pieces be foremost, while still informed by these mathematical ideas:

> I find it helpful to distinguish between compositions that are, in some sense, intended to be mathematical structures, and those that are motivated by mathematical structures. Both scenarios are creatively interesting to me and I strive to balance intense immersion in areas of mathematics, in collaboration with mathematicians, with periods when I allow my imagination to run wild.

Given our common interest in mathematics and music, Emily and I have set up a new centre at the Royal Northern College of Music in Manchester, where she is a professor, to explore the dialogue between the two disciplines. The centre is called PRiSM, which stands for Practice and Research in Science and Music. I like to think it's a modern-day version of the Corresponding Society of the Musical Sciences that Bach joined. Just as Bach's society brought together mathematicians, scientists and musicians interested in sharing ideas in the eighteenth century, we hope that PRiSM will do the same for practitioners in the twenty-first.

Hyperbolic coral reef

The hyperbolic spaces that Emily has been exploring are potentially the best geometric fit for how our universe is put together. One theory is that our universe looks like a dodecahedron with faces joined in the manner of a Seifert–Weber space. You can almost hear Plato saying, 'I told you so', although he hadn't conceived of the idea that the faces joined across the shape. Emily's *Compass* allows us to journey musically through this cosmos and experience something of that moment of passing through a face and reappearing twisted back inside the dodecahedron. Borges would have loved the idea too, as an extension of his Library of Babel built out of hexagons.

But you don't have to go to these cosmic limits to experience the beauty of hyperbolic space because nature is creating these geometries all around us on a more human scale. When things grow, the algorithm can quite often produce shapes that, instead of being flat, bend in two different directions in a way that resembles the roof of Hadid's London Aquatics Centre.

One intriguing way of seeing why algorithms for growth might produce hyperbolic geometries is by learning to crochet. It is a very mathematical craft: by interlocking loops of yarn or thread one at a time, a craftsperson can construct the most extraordinary variety of shapes. At one level, it looks very simple – you're just adding another small stitch. And yet, with a sophisticated enough algorithm, these simple additions can create a vast range of geometries. It's very close to one of the many ideas that Riemann introduced in his revolutionary work on geometry. What are now called Riemannian manifolds are shapes that locally look like flat pieces of Euclidean space, but sew them together and you can get exotic twisting geometries. It's exactly what the crochet artist is doing.

For example, if you wanted to crochet a ball, one way you could do it is to start by crocheting a ring of stitches to form the ball's equator and then gradually reduce the number of stitches in each subsequent layer you added on either side of the central ring. As you did this, the shape would gradually close off until you reached the poles of the sphere. The art – or mathematics – is deciding how many stitches to reduce each layer by so that you don't end up with a woolly rugby ball rather than a sphere.

But what would happen if, instead of decreasing the number of stitches, you increased them as you moved away from the central ring?

Now the crocheted fabric would bend in a way that accommodated this outsized addition, and as you kept going it would grow into a new sort of shape. The material would be curving away from itself. The result would be a crocheted piece of hyperbolic geometry.

When twin sisters Christine and Margaret Wertheim first tried crocheting shapes like this, they were struck by how similar they looked to forms that grow in nature, such as kale leaves. But the shapes that most resonated with them were those that they saw in the coral reefs just off the coast of Australia, where they were brought up. They both live now in California. Christine is an artist; Margaret a science writer. Their work on the crocheted hyperbolic geometries has brought together their different interests.

Their journey began in 2005 and was a response to the devastation that rising sea temperatures were causing to the corals that make up the Great Barrier Reef. The sisters came up with the idea of constructing an artificial reef, based on the fact that corals, kelps and various sea creatures are biological incarnations of these same hyperbolic geometries. One of the reasons that coral has evolved to grow in this way is that the curvy surface maximises nutrient intake. The geometry also ensures organisms are neither too rigid nor too flexible and allows them to move and change shape efficiently. The corals that became hyperbolic thrived . . . that is, until humans started interfering. The crochet coral reef that the Wertheims started to craft was their way of raising awareness of the devastation that our species was wreaking on this beautiful underwater geometric garden.

The crochet coral reef has continued to grow since those first pieces were made in 2005. I experienced the reef in 2008 when I took part in a workshop day organised to accompany an exhibition of the Wertheims' craft at the Hayward Gallery in London. The workshop involved learning both to crochet and to navigate the hard-core mathematics of hyperbolic space. Using the reef as an inspiration it explored the crossover between craft and science more generally. Today over 50 crocheted reefs exist around the world, crafted by 25,000 crocheters, or 'reefers' as they are referred to. The largest of these reefs – made up of 40,000 pieces crocheted by 4,000 contributors – can be found in Baden-Baden in Germany, displayed in what the Wertheims refer to as the Sistine Chapel of crocheted coral reefs.

The range of different hyperbolic shapes that have grown out of this simple idea of increasing the number of stitches in each row of crochet has given rise to some interesting new mathematical insights. Dr Daina Taimina, a mathematician at Cornell University, discovered a simple hyperbolic crochet surface which acts as a seed for the huge number of complex shapes that form these crocheted reefs. She had started experimenting herself with crafting these hyperbolic shapes after seeing examples that had been made from paper. But these were very fragile and prone to falling to pieces. She then tried to knit more robust versions, but they all turned out too floppy. But eventually Taimina found that crochet was rigid enough to make interesting shapes.

She pioneered some of the early crochet algorithms that so beautifully give rise to these hyperbolic geometries: crochet N stitches, and then on the next stitch crochet two stitches into one. Repeat this row after row and pure negatively curved geometry will emerge. Different choices of N determine how fast the curvature acts on the shape. Large N produces gently curving shapes that are almost flat but not quite. Small N and you get a shape that is turning in on itself, fighting to fill three-dimensional space. 'You can experiment with different ratios,' Taimina has advised, 'but not in the same model. You will get a hyperbolic plane only if you increase the number of stitches in the same ratio all the time.'

It was Taimina's development of these crochet algorithms that inspired the Wertheims to create the crochet coral reefs that are spreading across the globe. While Margaret's scientific upbringing conditioned her to follow the rules of the algorithm, Christine's instinct was to break them. The shapes that emerged by disrupting the numbers may not have resulted in pure mathematical geometries, but they have ended up looking more like the corals you see in the ocean. The random perturbations of nature disrupt the best algorithmic plans, so that growth is more likely to mimic Christine's subversive approach to crochet.

The Wertheims are very excited by this power of craftwork to explore fundamental science: 'The art-making *itself* recapitulates processes at the heart of organic evolution. Art becomes a tool not just for learning *about* science, but for *enacting* methodologies within nature that science uncovers.'

'There's no question,' Margaret Wertheim commented when she was in London for the workshop in the South Bank, 'that in the hierarchy of our society, science is up here and girls crocheting is down there. For many of the women who come, it's immensely uplifting to hear women's work "legitimated", as it were, by the contextualisation with science and mathematics.'

A metaphor for change

Although it is quite challenging to identify a novel whose narrative shape is non-Euclidean, there is no question that these new geometries which appeared in the nineteenth century provided a powerful mathematical metaphor for the way that society needed to question the blueprints it used to frame its current view of itself. Much to my surprise, I encountered a fascinating example of this when reading *The Brothers Karamazov.* I'd embarked on the book expecting another of Dostoevsky's grim tales of crime and punishment, this time in the shape of the apparent parricide by one of the brothers. But about a third of the way through the book, I was suddenly shocked to read a whole passage about non-Euclidean geometry as a metaphor for challenging the political status quo in Russia at the time.

Not only was Dostoevsky a great writer, he was also fascinated by the science of his day. He'd started his studies not in literature but in engineering in St Petersburg, where he was a student from 1838 to 1843. During his education, it seems that he struggled with algebra, but consistently excelled in geometry. So he would have been well equipped to appreciate the new ideas that emerged during this period concerning geometries that challenged Euclid's blueprint of the universe.

Significantly, Dostoevsky was excited not only because a successful challenge had been mounted to the old axioms on which geometry was based, but because a Russian mathematician, Nikolai Lobachevsky, had been one of those to lead it. Russia was considered something of an intellectual backwater at the time – Dostoevsky's engineering school had been founded only in 1810 as part of an attempt to keep pace with the scientific advances of other European powers – so seeing one of their own at the head of this mathematical revolution was something the Russian intelligentsia was keen to celebrate.

The idea of new blueprints for geometry chimed perfectly with Dostoevsky's wish to explore in his novel a character who questioned the existence of God. Listen to Ivan Karamazov probing the hypotheses on which the society he inhabited was established:

> If God exists and if He really did create the world, then, as we all know, He created it according to the geometry of Euclid and the human mind with the conception of only three dimensions in space. Yet there have been and still are geometricians and philosophers, and even some of the most distinguished, who doubt whether the whole universe, or to speak more widely the whole of being, was only created in Euclid's geometry; they even dare to dream that two parallel lines, which according to Euclid can never meet on earth, may meet somewhere in infinity.

Albert Einstein was very taken by these literary explorations of the geometries that he depended on for the construction of his theory of relativity. 'If you ask in whom I am most interested at present, I must answer Dostoevsky – Dostoevsky gives me more than any scientist, more than Gauss!'

Dostoevsky's novels resonated so strongly with many of their early readers because they heralded the ideas that would emerge in the century to follow. His view of art, for example, was very dismissive of realism. He wrote once about visiting an exhibition of the hyper-realist Peredvizhniki School and finding the paintings hollow, stupid and superficial. For him the truth lay deeper, in the abstract structures that underpinned reality. It was this that artists such as Malevich would subsequently strive to capture and which eventually would inspire the buildings that Hadid has created in London and across the world.

Hadid returns

Given Hadid's affinity with mathematics, it was perhaps entirely appropriate that she should have been chosen to design the new mathematics wing of the Science Museum in London. It opened in 2016, just a few months after Hadid died at the age of 65. The centrepiece of the gallery is a Handley Page biplane suspended from the ceiling, one of the first experimental aircraft that ushered in the age of commercial aviation. Hadid's design was inspired by the mathematics describing turbulence behind the wings of the plane. Fully understanding this turbulence is one of the greatest unsolved challenges of mathematics today. It requires solving some of the most elusive equations on mathematicians' books: the Navier–Stokes equations.

Hadid took the shape of the vortices in the turbulence created behind the plane in flight and used them to define the gallery space. These shapes are governed by six different parameters; dialling these up and down causes a menagerie of sensuous surfaces to appear. This mathematical approach to creating interesting forms has spawned a whole new movement in architecture called 'Parametricism', which Hadid's team in London pioneered. The name refers to this process of representing something with parameters and then varying the parameters to explore the range of different possibilities. This idea of Parametrisation, with its infinite possibilities, raises the prospect of designing buildings that are absolutely unique – a concept of architecture far removed from Le Corbusier's dream of standardisation.

The exciting thing about Parametricism is the astonishing new vistas it opens up. The mathematics can produce surprising forms that perhaps would have been beyond the human imagination to conceive. As Hadid said:

> Mathematics and geometry has an amazing influence on our work. It's very exciting. Huge advances in design technology are enabling us to rethink form and space, and the wonderfully fluid surfaces and structures of each project are defined by scientific innovations.

It is this beautiful fusion of mathematics and art that visitors see on display as they explore the mathematical exhibits in the Science Museum.

But although the computer is important in realising the ideas, the art is in creating the mathematical relationships that become the code that is fed into the computer. And those mathematical relationships will very often start with a simple hand-drawn sketch. As Hadid said, mathematics for her was always a form of sketching. Many people believe that the computer has similarly put mathematicians out of business, but this is to misunderstand mathematics. The computer is a useful tool for exploring mathematical ideas, but it can't come up with the original concept (or at least not yet).

Hadid's work is the fusion of mathematics and art at its best. As Le Corbusier wrote: 'For the artist, mathematics . . . is not necessarily a matter of calculation but rather of the presence of a sovereign power; a law of infinite resonance, consonance, organisation.' The mathematics allows the artist access to new possibilities that could never have been envisaged without the logical constraints of the equations that propel the artist through this narrow passage into a landscape of new forms. But for me the real excitement comes from being able to see, touch and inhabit abstract mathematical shapes in my neighbourhood of London that were once just the preserve of the mathematician's mind.

My final blueprint might better be described as an anti-blueprint. Nevertheless, chance has been an important element shaping the evolution of the universe and the creative practice of many artists. But it is only with mathematics that we have been able to tame randomness.

Blueprint Nine
Randomness

Jean Arp had been working on a drawing for some time, but dissatisfied with his efforts he ripped it to pieces and threw them in the air. Arp was desperate to try something new. His artistic training at schools in France and Germany in the first decade of the twentieth century had already led to his work being shown alongside that of Matisse and Kandinsky. But the year was 1916 and Arp had been profoundly affected by the horrors of the First World War which were being played out around him. It needed a new artistic vision to respond to the trauma of modern warfare which his generation was witnessing. He was becoming increasingly discouraged with his efforts. Nothing was working. But when he returned some hours later to tidy up his mess, he was transfixed by what he saw.

As his friend Hans Richter recalled about Arp's reaction to the image that lay on the ground:

> It had all the expressive power that he had tried in vain to achieve. How meaningful! How telling! Chance movements of his hand and of the fluttering scraps of paper had achieved what all his efforts had failed to achieve, namely expression.

Arp carefully pasted down the paper shreds just as they had fallen on the floor: the first in a lifelong series of collages created, as he acknowledged, 'according to the laws of chance'. 'Chance is my raw material,' he was later to declare.

It might seem strange to celebrate randomness as a blueprint. After all, randomness feels like the opposite of structure. It is an anti-blueprint. A disrupter of expectation and pattern. But randomness has been a powerful agent forcing artists to step outside their comfort zone and confront the new. Just as Arp found that randomness can lead you into realms that the conscious mind might have been blocking.

The way those pieces of paper settled on the floor sparked Dada, one of the most significant art movements of the early twentieth century. Emerging from the trauma of the First World War, the Dada movement was as much driven by politics as it was by trying to find a new artistic language.

In the Dadaists' eyes, the war was the culmination of an obsession with rationalism and logic, with aestheticism and capitalism. The aim of Dada was to disrupt all this, to upset norms and expectations, to offend. 'We had lost confidence in our culture,' explained the Romanian artist Marcel Janco, one of the driving forces behind the movement. 'Everything had to be demolished. We would begin again after the tabula rasa.' As Marcel Duchamp characterised it, Dada was anti-art intended to overturn the accepted definitions of what art was – as he himself famously achieved in 1917 by exhibiting an ordinary urinal in a gallery, under the title *Fountain*. The discovery of the disruptive power of chance led Dada to adopt randomness as an important weapon in its artistic arsenal. According to Richter: 'Chance must be recognised as a new stimulus to artistic creation.'

The Cabaret Voltaire, an avant-garde nightclub in Zurich, became the stage, both literally and figuratively, for much of their first creative output. 'At the Cabaret Voltaire,' Janco recalled, 'we began by shocking common sense, public opinion, education, institutions, museums, good taste, in short, the whole prevailing order.' The reaction of one art critic was visceral: 'Dada philosophy is the sickest, most paralyzing and most destructive thing that has ever originated from the brain of man.' This was exactly the response Dada wanted.

The reason for the name 'Dada' has never been definitively pinned down. True to the role of chance in the act of creation, it could well have been chosen randomly. One theory goes that a knife was inserted in a dictionary at the French word for a hobbyhorse: *dada*.

There is some question whether Arp truly submitted to chance as these pieces of paper fluttered to the floor, or whether he couldn't resist rearranging them into a more pleasing configuration. The artwork that emerged from that first experiment has the roughly square pieces in quite a regular grid formation. There are none of the overlaps that would be expected of pieces falling to the floor. But Arp always claimed that chance was just an ingredient, not the author of the piece. It was meant to propel him into unanticipated territory and allow for the discovery of new structures that his conscious mind had inhibited.

For Arp, there was an important distinction between chance and randomness: 'In the gibberish with which people happily augment the confusion, the word chance is used very much like senselessness. I have never called my works "according to the law of senselessness".' Chance was part of the natural order and hence incorporating it into his creative process was consistent with an artist's response to the universe around him. It seems that Arp felt there was something more authentic, maybe even romantic, in the word 'chance' than existed in the idea of randomness.

The use of randomness or chance to inspire creativity in the visual realm did not originate with Dada. Leonardo da Vinci had suggested hurling a paint-soaked sponge at a wall and used the resulting marks as the basis for his paintings. In the compilation of his observations published after his death as *A Treatise on Painting*, Leonardo advised artists to explore the random marks in the world around them:

> If you look upon an old wall covered with dirt, or the odd appearance of some streaked stones, you may discover several things like landscapes, battles, clouds, uncommon attitudes, humorous faces, draperies, etc. Out of this confused mass of objects, the mind will be furnished with an abundance of designs and subjects perfectly new.

Francis Bacon admitted to taking Leonardo's advice: 'My ideal would really be just to pick up a handful of paint and throw it at the canvas and hope that the portrait was there.'

The power of chance to stimulate creativity is all down to the human mind being primed to find a story lurking even in a random blob. This urge to find meaning is at the heart of the famous inkblots Hermann Rorschach used in his work as a psychoanalyst. The added symmetry that Rorschach introduced contributes to the brain's belief in a hidden meaning because, as we saw in the blueprint dedicated to symmetry, its occurrence in nature often encodes meaning or a message. Rorschach believed that what you saw in these random symmetrical blobs revealed something about the patient's state of mind. Introverts emphasised movement in the images, grumblers and pedants were obsessed with small details in them, while those shocked at the first appearance of colour in the blots were diagnosed as repressed.

You might wonder why I didn't choose the paintings of Jackson Pollock as the perfect example of the role of chance in art. Although there is a semblance of randomness in them, Pollock was very clear that he was totally in command of the technique and the actions he performed to create them. This reveals an interesting tension between deterministic chaos and random chance. The roll of a dice might be determined by the conditions as it leaves your hand, but the chaotic nature of the equations that control the outcome means that very small differences in this starting data result in different scores. Because of our inability to manage the chaos, we instead resort to the language and mathematics of chance to analyse the throw of the dice. This is subjective randomness, where our ignorance of the underlying mechanism determining the outcome means that we resort to calling it chance.

The existence of objective chance is at the heart of a heated debate in modern physics. Until the advent of quantum physics, it was believed that the universe was like a piece of clockwork, ticking along to the laws of physics that Newton had discovered. We may not know the precise set-up of the universe which would be needed in order to access the predictive power of the equations; but if we did, then, as the eighteenth-century scientist Pierre-Simon Laplace famously articulated, we could know the future.

But the physics of the early twentieth century offered a challenge to this deterministic view of the universe. The future path of a particle was not determined by its history. There is no physical model that will predict when a piece of uranium will emit its next bit of radiation. The current model of physics has genuine randomness embedded at its heart.

What is fascinating is to see the art of the early twentieth century beginning to reflect the role of chance in creating reality. As Umberto Eco wrote: 'In every century the way that artistic forms are structured reflects the way in which science or contemporary culture views reality.'

Random writing and reading

Cutting up your material and scattering it on the floor to stimulate new ideas was not just the preserve of the visual artist. David Bowie used a similar idea to write lyrics for his songs. He was inspired by William Burroughs, who in the 1960s developed what he called the 'cut-up technique'. 'Cut-up the words and see how they fall,' Burroughs said. Burroughs believed that using randomness to reorder old lines of poetry would instil new life in the text:

> Take any poet or writer you fancy ... or poems you have read many times. The words have lost meaning and life through years of repetition. Now take the poem and type out selected passages. Fill a page with excerpts. Now cut the page. You have a new poem. As many poems as you like.

Bowie so enjoyed the technique that he got a friend to write some software so that he could use his computer to do the cutting up without him having to get out a pair of scissors: 'I'll take articles out of newspapers, poems I've written, pieces of other people's books, and put them all into this little warehouse of information and hit the button, and it will randomise everything.' He first featured cut-ups on his 1974 album *Diamond Dogs*, but continued using the idea right through to his final two albums, *The Next Day* and *Blackstar*.

Burroughs's idea of using randomness as a way to reinvigorate old works that had lost their meaning through being heard too many times was part of the motivation behind several of Oulipo's projects. N+7 was meant to be applied to texts we knew well, but the replacement of the nouns in the poem with the ones seven words further along in the dictionary made us hear the writing afresh.

Queneau, the founder of Oulipo, encouraged the use of randomness in the way that he made you choose which of the one hundred thousand billion poems you would read by picking lines from the initial ten sonnets that he wrote. When *A Hundred Thousand Billion Poems* was published, the ten sonnets were bound together, but the pages were cut between each of the lines of the poems in such a way that you could pick the lines you wanted simply by folding the others over.

Many writers in the twentieth century have experimented with the role that randomness can play in creating literature, not only by the author but also in how a novel might be read. Published in 1963, Julio Cortázar's *Hopscotch* consists of 155 chapters. The first 56 are divided into two narratives called 'From the Other Side' and 'From This Side'. The first tells the story of Horacio Oliviera, an Argentinian living in Paris. The second narrates the tale of his childhood friend Manolo Traveler and is set in Buenos Aires. The final 99 chapters are designated as 'expendable' and come under the heading of 'From Diverse Sides'.

You can read the novel straight through, but Cortázar suggested an alternative way of reading it which jumps through the narrative as if playing hopscotch, following a route mapped out via instructions at the end of each chapter. The result is a potentially infinite book because it ends with you jumping back and forth between Chapters 58 and 131. Although Cortázar indicated that this was just one choice of many that the reader could make. The reader was also free to choose a random path through the chapters.

The invitation to jump randomly through the narrative is meant to mimic life. Cortázar was trying to capture the haphazardness that seems to make up our journey from birth to death. As Horacio explains in the first part of the novel: 'I imposed the false order that hides chaos, pretending that I was dedicated to a profound existence while all the time it was the one that barely dipped its toe into the terrible waters.'

As well as the array of human characters who populate the book, there is also a rather odd circus cat who makes an appearance in the chapters set in Buenos Aires. It is one of my favourite characters in the book. The cat's claim to fame is that it can count and do calculations. Perhaps it isn't surprising to learn that Cortázar had a cat himself called Theodor W. Adorno, named after the German philosopher. The cat features extensively in a collage-like collection of short pieces Cortázar wrote called *Around the Day in Eighty Worlds*.

I wonder, though, whether Cortázar's approach in *Hopscotch* works. I read the novel straight through in one go and then I glanced at the alternative he suggested but I'm unlikely to read it many more times in my life. I'm sure my experience is not unique, so I would question how many times it will genuinely have been read randomly as Cortázar also offers.

B. S. Johnson is another author who used chance to create multiple ways of reading a narrative. In a similar manner to Cortázar, he wanted the text of his novel *The Unfortunates* to represent the fragile nature of memory as well as the randomness of loss and illness. But, in some ways, I think it is much more successful as a project. Published in 1969, it consists of 27 chapters collected together in a box. The first and last chapters are determined, but after that the reader can assemble the other 25 chapters in any order they wish, in effect creating their own unique book.

The physical construction of the book as 27 disconnected chapters left unbound in the box encourages the reader to assemble the chapters into a random order. Perhaps Johnson's work is more successful than Cortázar's *Hopscotch* precisely because he gives up complete control of the text. Cortázar's suggestion for an order to read the passages results in a bias in the possible choices. The fact that Cortázar's book is bound is also working against the randomisation.

The Unfortunates recounts Johnson's own experience of being sent as a sports journalist to a city to report on a football match, only to realise that the city had once been home to a friend who'd died of cancer. The visit prompts memories of this friend from their shared student days to his funeral, interspersed with episodes from his day spent in the city, watching football ('City one, United nil'), eating chips ('deep in my heart I know I love chips') and drinking marsala wine at Yates ('one more Marsala, no, I'll fall asleep this afternoon').

I hadn't heard about *The Unfortunates* until I read *Like a Fiery Elephant*, Jonathan Coe's wonderful biography of Johnson. I was a judge in 2005 for the BBC Four Samuel Johnson Prize for Non-Fiction, and Coe's book was the worthy winner that year. Reading about how Johnson had used randomness to frame *The Unfortunates* whetted my appetite to see how effective the idea had been, or whether the randomness destroyed the narrative. But the book proved notoriously difficult to track down.

My first experience of the book had to wait until 2011, when I heard a wonderful radio adaptation by the BBC, with Martin Freeman playing the narrator, which genuinely randomised the experience of the story. And it really worked. The way one jumped around from the memory of eating lychees together to the news of his friend's death to the report of the football match felt like a fragmented but very honest representation of how our brains tell stories. The adaptation broke the novel up into 17 sections, and the BBC offered listeners the opportunity to hear the sections in any order, exploiting some clever software developed for the Alexa device.

I was still keen to get my hands on a physical copy of the book. And then I had an idea. My library in Oxford, the Bodleian, was the first legal deposit library in the UK. Sir Thomas Bodley's agreement with the Stationers' Company that any newly published book must be added to the collection in Oxford was formalised in law in 1710. The publishers of *The Unfortunates* were legally obliged to deposit a copy of the edition I was after in the library, so I ordered it up.

I sat for a wonderful morning in the Radcliffe Camera, the part of the Bodleian Library designed by architect James Gibb, constructing my own journey through Johnson's memories. The first and last chapters were fixed but the other 25 sections that could be read in any order, rather than having a chapter heading or number, came with their own little symmetrical icons running like a frieze pattern at the top of the first page. It was exciting to think that this was one of 15 million billion billion possible ways of reading the story, which meant that mine was almost certainly the first time it had ever been read in this order. Although at first sight the book smacks of being avant-garde and unapproachable, the opposite is true. It is full of humour, pathos and humanity, where the structural conceit supports perfectly the exploration of the nature of memory.

Johnson's approach confirms Eco's idea that art mimics the prevailing scientific philosophies of the age. As Johnson articulated:

> Present-day reality is markedly different from say nineteenth-century reality. Then it was possible to believe in pattern and eternity, but today what characterises our reality is the probability that chaos is the most likely explanation; while at the same time recognising that even to seek an explanation represents a denial of chaos.

Johnson hoped that the uncertainty created by the unbound chapters would embody in physical form the contradictory and unstable anxieties described in the novel. But not everyone was a fan of allowing chance a role in the creative process. Samuel Beckett, for example, was critical. Although Beckett had experimented with chance in his prose piece 'Lessness', he believed ultimately that randomness in literature destroyed the possibility of meaning. Johnson, however, felt it might help the reader to find meaning.

Our desire for order in the randomness means that, just as with Rorschach's inkblots, we find meaning in whatever path we take. As Johnson wrote: 'While I believe (as far as I believe anything) that there may be (how can I know?) chaos underlying it all, another paradox is that I still go on behaving as though pattern could exist.'

One of the aims of allowing a random reading of a text is to try to involve the reader as an active participant in the creation of the story.

It is the same technique that is used in the 'Choose Your Own Adventure' series of books, where you are asked to decide which direction to take the story. 'Turn to page 35 if you want to go into the cellar, page 47 if you want to climb to the attic.' Often these books encourage agency by giving you context for your decisions, while a truly random approach to the choice denies you that agency precisely because the author wants to disrupt your normal mode of living.

There is a difference between using chance as an ingredient in your creative process as an artist and creating a piece where chance generates a new piece every time it is engaged with. If you are using chance to stimulate ideas then you still have the ability as the artist to reject or adapt what chance has offered you, although some artists are against any sort of meddling. If you are using chance in your work, then I think you should let it have its voice. Arp very probably rearranged the torn pieces to make a more satisfying configuration for that first collage created according to the laws of chance. But if you are intervening, then in my view it weakens the role that randomness is playing. That said, the latter approach can still take you in an unexplored new direction.

But then there is the artist who allows the viewer or listener or reader to make the choices themselves. This is a very different process. You have to create a piece that will work in multiple ways, often at a scale which means that humanity will never generate all the possible iterations of the piece. The skill of the artist then lies in making a machine which allows for interesting and varied pieces that feel like they work no matter what direction chance takes you in. Some have criticised those artists who use chance in this way for abdicating their responsibility for the creative process, but for me it acknowledges that the viewer or listener as they engage with a work is always part of that process.

Aleatoric music

The huge surge of work in the twentieth century that started using chance in the creative process led to a new term being created to describe this practice: 'aleatoric art'. The name has its origins in the Latin word for dice: *alea*. The term first caught on when applied to the role of chance in musical composition rather than visual art or literature.

It was championed by Pierre Boulez, who first saw the term used by Werner Meyer-Eppler, a German physicist specialising in acoustics. Meyer-Eppler's work on electronic music led him to the idea of sound that is 'determined in general but depends on chance in detail'. He named this concept with the German noun *Aleatorik*, but as correspondent Arthur Jacobs of London N20 pointed out in a letter to the editor of the *Musical Times* in May 1966 the English version mistranslated it into the adjective 'aleatoric'. 'May I hope that, as a result of this exhortatoric letter, this unsatisfactoric (and to me inflammatoric) usage will prove transitoric?' It didn't and the word aleatoric is now used to describe works of art that involve chance.

A watershed in the way that chance was used in musical composition occurred with John Cage's *Music of Changes*. The complete solo piano piece was first performed on 1 January 1952 by Cage's friend David Tudor. Cage commented that Tudor had to learn 'a form of mathematics which he didn't know before . . . a very difficult process and very confusing for him.' Mathematics had always been part of Cage's world but he had wrestled with how to incorporate it in his music. Asked what his first music was like, he responded:

> It was mathematical. I tried to find a new way of putting sounds together. Unfortunately, I don't have either the sketches or any clearer idea about the music than that. The results were so unmusical, from my then point of view, that I threw them away.

The challenge for any pianist performing Cage's *Music of Changes* is that the music constantly shifts in tempo, speeding up one moment, slowing down at others. Piecing this all together into a coherent performance requires a level of mathematics, on top of the musicology, to make sense of what Cage was asking. But it is the way that Cage made these decisions about tempos that is interesting, because he used chance as embodied in the ancient Chinese practice of divination using the I Ching, otherwise known as the 'Book of Changes', hence the name of the piece.

Cage had become hugely influenced by the ideas of Zen Buddhism, which led to his encountering the I Ching. This book is used in a system of divination involving sticks, which are used to randomly determine a hexagram, a symbol made up of a combination of six solid or broken lines. There are $2^6=64$ different symbols possible, and each symbol corresponds to a different text in the I Ching which is then used as the basis for interpreting a divination.

Reading about these hexagrams in the I Ching is thought to have inspired Leibniz's ideas for representing numbers in binary. The broken lines can be thought of as 0s and the solid lines are 1s so that each hexagram corresponds to a number in binary.

Essentially Cage wanted chance to influence his choice of almost every aspect of the music: tempos and dynamics, silences and sounds, the duration and how many instruments can be heard at any one moment. This was an example of extreme serialism. Schoenberg had put the 12 notes of the scale on equal footing in the first move towards serialism. Composers of this period, including Boulez and Karlheinz Stockhausen, wanted to serialise everything.

Cage assigned each musical quality its own 8×8 chart, with every entry corresponding both to some decision about the sound and also a specific hexagram from the I Ching. Traditionally, each hexagram was determined by a complicated process of passing sticks from one hand to the other, but Cage instead used a coin. The coin would be tossed six times. Tails corresponded to a broken line in the hexagram, heads to a solid line. The number of decisions that had to be taken to compose the piece meant a huge amount of coin-tossing. Cage ended up employing friends to toss coins whenever they came to visit.

Although Cage's charts were indexed and accessed using the symbols of the I Ching, it doesn't appear that there was necessarily any correlation between the traditional meanings of the hexagrams and the musical values that Cage assigned to them. The connection to Zen Buddhism feels rather tenuous; it merely inspired the way that Cage was deciding on the elements of his music. He was keen to stress that although chance was used to decide the micro-structure of the piece, the content of the charts was under his control as a composer: 'These charts were subjected to rational control . . . the effect of the chance operations on the structure (making very apparent its anachronistic character) was balanced by a control of the materials.'

One aspect of structure that was very much in Cage's control was the way the piece was divided into units. This reveals an interesting fractal blueprint that Cage loved to use in his music, where the macro-structure of a piece is reflected in the micro-structure. *Music of Changes* is divided into 29 units, which are grouped in the proportions 3, 5, 6¾, 6¾, 5, 3⅛. This large-scale structure is then repeated at the smaller scale, where each unit in turn is divided in the same proportions.

The result is indeed extraordinary. As David Ryan wrote in notes to accompany a recording of the piece by the pianist Tania Caroline Chen: 'It is music of great austerity and beauty – glacial and impassive, and yet requiring virtuosic concentration and execution.'

Cage had been trying to find a way to remove himself from the compositional process. He wanted to liberate himself from his own personal tastes and memories as well as to distance the work from previous musical styles and traditions. The element of chance was his tool for achieving this separation. It was also, for Cage, a response to nature. He believed 'the responsibility of the artist is to imitate nature in her manner of operations'. Once again, we see the common thread of mathematical blueprints used in art simply being the blueprints that we also find underpinning the natural world.

Looking back at *Music of Changes* some years later, Cage felt that the role that chance had played rendered the work more inhuman than human. The way it controlled the performer gives the work an alarming aspect of a Frankenstein monster. The situation is of course characteristic of Western music, the masterpieces of which are its most frightening examples, which when concerned with humane communication only move over from Frankenstein to Dictator.

Choose your own adventure

Once Cage had used the tossing of a coin to compose *Music of Changes*, from that point on it was frozen in time and tied down. But other composers, such as Boulez and Stockhausen, started experimenting with the idea of pieces that allowed the players to choose a pathway through the composition. This fitted perfectly with Meyer-Eppler's notion of aleatoric music, in which the composer made the global decisions but chance provided localised variations.

Boulez's piano composition 'Constellation-miroir' took as its inspiration a famous poem by the late nineteenth-century French poet Stéphane Mallarmé. Intriguingly called 'A Throw of the Dice Will Never Abolish Chance', it exploits typographical layout as an important ingredient in writing. The sparse text meanders back and forth across the page following an irregular path that makes how it should be read ambiguous. There are huge swathes of empty space and the typography is constantly changing. No dice were used in the writing of the poem; the title refers to the dice that are in the hand of the Master who in the poem is about to be drowned in a shipwreck. He hesitates to throw them into the waves, convinced that they hold the secret to an important number. French philosopher Quentin Meillassoux has tried to identify whether this number might be encoded somehow in the text itself in a monograph he published in 2011. In *The Number and the Siren* he argues that the numbers 7, 12 and 707 feature prominently in the poem's composition.

The poem was an inspiration not just for Boulez but also for the Dada movement and would often be recited at their soirées in the

Cabaret Voltaire. Boulez was especially taken by the layout, which hinted at ways to chart a path through the lines. The score of 'Constellation-miroir' captures something of the visual quality of Mallarmé's poem. The 58 segments are laid out across nine pages printed in red or green. Red indicates a sequence of several chords; green denotes passages of single notes. The possible routes through the piece and the various ways to play each segment are indicated at the beginning and end of each part.

Although 'Constellation-miroir' is hailed as an aleatoric composition, in which the pianist can choose their own route through the score, in truth Boulez loaded the dice. The piece gives you the illusion of choice as a player: as you explore the conditions and restrictions that are in the score you realise Boulez ceded control of very little – and, frankly, what he did give up has little impact on the performance.

Karlheinz Stockhausen, in contrast, wrote a piece for pianists that allowed a lot more freedom of choice. Stockhausen had studied under Meyer-Eppler at the University of Bonn, and his ideas of the *Aleatorik* might well have influenced the direction that Stockhausen took in the composition of his *Klavierstück XI*.

The piece consists of 19 musical fragments which are spread across a single page of score. The performer is allowed to start with any fragment; then, at the end of that fragment, they find instructions which control the tempo, dynamics and articulation for the next one. They are told how to play that next fragment, but it's up to them to select which one it will be. The piece continues with the pianist choosing one fragment after another. They are allowed to repeat a fragment they have played already, but if it is preceded by a different fragment this time around, then it won't be performed the same way as before because it's governed by that second fragment's instructions. They can also choose to alter the pitch the second time, for example by playing it up an octave. The piece ends when a fragment is performed for a third time; there is no requirement to play each and every one of them.

Stockhausen's idea was that when a player performed any piece by any composer, then the performance would be different every time. The performer would interpret in their own unique way how the notes were played, both individually and collectively. Listen to multiple recordings of a piece such as Beethoven's 'Moonlight' Sonata and you will hear they vary immensely. Different dynamics, tempos and choices of articulation. Stockhausen was just taking interpretation at the micro-level of the piece and applying it to the macro-structure.

Stockhausen perceived *Klavierstück XI* as resembling a single vibrating molecule made up of 19 atomic components. These fragments were also composed using a very mathematical structure that involved a system of matrices or rectangular arrays of numbers to assign rhythm and pitch to each musical section. Stockhausen was someone who took serialism to its logical conclusion. Just as Schoenberg had used permutations of the 12 notes of the chromatic scale, these matrices were permuting rhythms and dynamics and other musical qualities.

I would question whether a piece such as *Klavierstück XI* is truly aleatoric. The performer is making choices about the order in which passages are played rather than giving up the decision to pure chance. Given that they would have been practising the fragments, it seems inevitable that personal preferences would draw them to certain parts – and combinations of parts – more than others. Even if the eye were allowed to roam freely over the score in order to pick the next fragment, there are often patterns to how we scan a page. The performance would inevitably reflect more of the performer's character than truly aleatoric music would allow.

The question of whether humans can truly pick randomly is a fascinating one. We are pattern-searchers and pattern-makers. We find it very hard to behave genuinely randomly. To truly create aleatoric music, the trick is to exploit chance generated by external forces outside human control, such as the toss of a coin or the throw of a dice.

Composing by dice

The musicians of the twentieth century advocating for the use of chance in composition may not have been as radical as they thought. A number of pieces composed in the eighteenth century use a set of dice to determine the notes that are played. One of the first was created by Johann Philipp Kirnberger, who was a contemporary of Bach and trained with the great master for some years. Devised in 1757, it was called 'The Ever-Ready Minuet and Polonaise Composer'.

The game was designed to generate either a 14-bar polonaise or a 32-bar minuet and trio. Kirnberger composed 11 different options for each bar of either piece. The novice composer would then simply roll two dice, giving a score of 2 to 12, which was then used to pick the bar to be played. The aim was to allow the musically untrained to feel that they were composing. As Kirnberger explained: 'Anyone who is familiar only with dice and numbers and can write down notes is capable of composing as many of the aforesaid little pieces as he desires.'

At the front of the booklet in which the game was printed, were two large tables of numbers: one of 14 rows and 11 columns for composing polonaises; and one of 32 rows and 11 columns for the minuets and trios. The rows corresponded to the number of bars in each piece, and the columns to the roll of the two dice. The remaining 29 pages consisted of numbered bars of music. The player was required to cut out all of these bars and stack them up like a pack of cards. Then they would throw the dice, and consult the table to see which bar of music had been chosen. These were gradually pieced together until the composition was complete and ready to play.

Rather like Queneau's sonnets, Kirnberger had to compose 11 different polonaises and minuets in such a way that, when cut up and stitched together using the dice, the resulting pieces wouldn't sound too clunky. It seems that Kirnberger was unsure how many different pieces of music he'd ended up composing because he resorted to asking a mathematician friend to do the calculation. It's not difficult. With 11 choices for each of the 14 bars of the polonaise there are 11^{14} possible pieces. For the minuet and trio, it comes to 11^{32}. These numbers are so large that even if the dice had been rolled since the Big Bang and the polonaises played non-stop, we would only have got through about 2 per cent of all the possibilities by today.

One of the disadvantages of using two dice is that some scores are more likely than others. For example, getting a 7 with two dice is six times more likely than getting a 12. This is because there are six ways to score 7 and only one way to score 12. This inequality means that the pieces generated will be biased towards those bars numbered in the middle of the range of the dice rather than the extremes.

One composer who saw the drawback of using two dice was Bach's son Carl Philipp Emanuel Bach. His 'Invention by which Six Measures of Double Counterpoint Can Be Written without a Knowledge of the Rules' employed a nine-sided spinner to generate its harmonies. This mechanical composer controlled notes rather than bars and was really a dig at the music Carl Philipp Emanuel's father had been composing. The younger Bach had always been rather dismissive of the rather dry nature of the fugues that Johann Sebastian pumped out. He thought the whole thing very formulaic and declared himself 'no lover of dry mathematical stuff'. The 'Invention' was meant to demonstrate how mechanical the whole process was.

When the piece was published in 1757, it was accompanied by four pages of mathematical notes written by Friedrich Wilhelm Marpurg explaining the compositional tricks and probabilities involved. Marpurg was well qualified to explain the mathematics of chance and ended up being director of the Royal Prussian Lottery for thirty-two years. He calculated that the number of different pieces that C. P. E. Bach's machine could generate was a mere 282,429,536,481, which would take about half a century to perform back to back.

Kirnberger's musical dice game was clearly a hit as it inspired the creation of some 20 other such mechanical composers. One of them, which wrote waltzes, is attributed to Mozart. His 'Musikalisches

Würfelspiel', or Musical Dice Game, was published in 1792, a year after his death. But there has been some speculation by historians that the piece was created by his publisher Nikolaus Simrock and released under Mozart's name in a bid to boost sales.

The fascination for using dice in musical composition during this period reflected a growing interest among the public in mathematics and especially its power to navigate the world of chance. The pioneering work a century before by Pascal and Fermat on the mathematics of probability was becoming mainstream. This was an era when many composers were thinking in quite mathematical ways. The French composer Jean-Philippe Rameau had written a whole treatise in 1722 explaining how mathematics could be used in composition. Keyboard players were experimenting with different types of tuning informed by mathematical theories. Kirnberger's own writings were criticised by his English contemporary Charles Burney as being 'more ambitious of the character of an algebraist, than a musician of genius'. It further confirms Eco's idea that art very often reflects the science of its time.

The bird's nest and beyond

'Aleatoric architecture' would seem to be a contradiction in terms. Architecture is about controlling your environment, precision engineering, perfectly executed blueprints for the urban landscape we inhabit. There is little room for randomness in designing a building. But there have been architects who have tried to simulate a sense of it at play in their constructions while still maintaining control.

The National Stadium built for the 2008 Olympics in Beijing was dubbed 'the Bird's Nest' because the metal struts that covered its exterior looked so randomly arranged that it gave the impression of a giant nest built out of sticks. But this random-looking façade is pure decoration and not a structural part of the building itself. That was still designed and executed with nothing left to chance.

The concrete façade of the Barbican Centre in London has a completely random rough texture which was created by the builders taking pick-hammers to its surface. Due to the nature of the process, the pattern on the wall is never repeated. This random texture was not the original idea for the building. The architects had initially proposed smooth, polished surfaces, including marble cladding, but the City of London Corporation, which was responsible for paying for the project, decided this would be far too expensive. Taking a hammer to the building was the architects' act of protest at having their vision rejected, but it ended up as one of the Barbican's distinctive features. But like the National Stadium in Beijing, this was superficial randomness rather than something fundamentally structural.

There is, however, a movement that is beginning to challenge the idea that chance shouldn't play a part in the construction of buildings. Academics Sean Keller and Heinrich Jaeger have started to explore whether it might be possible to create aleatoric architecture. As they have argued, the micro-structure of a building is not determined. At a granular level, construction materials such as concrete are left to form part of the building according to the push and pull of the local environment. The architect does not have any control of them at this micro-level. Is it possible to amplify this lack of agency up to a larger-scale structure? The thinking is similar to Stockhausen's argument that because performers were already introducing aleatoric elements into compositions simply by the way they played them, why not allow randomness to determine the global structure too?

One way to do this with architecture would be to exploit the unpredictability of the granular composition of building materials to see whether it gives rise to interesting new structures as the granular components bind together to make a larger structure. Nature of course already does this, so it doesn't seem impossible to imagine the same approach being successful in human construction. The twentieth-century architect Louis Kahn always believed that a material knew what it wanted to be. You just had to ask it.

> You say to brick, 'What do you want, brick?' Brick says to you, 'I like an arch.' If you say to brick, 'Arches are expensive and I can use a concrete lintel over an opening. What do you think of that, brick?' Brick says: 'I like an arch.'

The hope is that modern materials might be more flexible than the stubborn brick.

Advances in technology have allowed a degree of freedom in architecture that has been impossible in the past. Computers can be programmed to know the parameters within which a building is still stable and constructible. But then the architect has the flexibility to play within these parameters of stability. As we saw when we discussed Hadid's legacy, this has led to a whole new movement in architecture called 'Parametricism'.

What is truly random?

Tossing a coin just as John Cage did to compose his *Music of Changes* seems as good a way as any to randomise choices. But how can we be sure that the outcome is truly random? For example, which of the following sequences of 20 coin-tosses would you feel has been rigged?

H T H T H T H T H T H T H T H T H T H T
or
H H H H H T H T H H H T T T T H H H T H.

Clearly the first sequence feels like it hasn't been generated in a random fashion because it has a very obvious pattern. But the second sequence might worry some of you because that is a long run of heads to be starting with and there are more heads than tails. Is this coin biased?

You might argue that the first sequence is highly unlikely. What are the chances of getting such a perfect sequence of heads and tails? The answer is 1 in 2^{20} which comes out at a bit more than 1 in a million. So, very unlikely. But that isn't a good measure of randomness because the probability of producing the second sequence is exactly the same. And the same goes for any other sequence of 20 coin tosses you could imagine. So probability is not a good test of randomness.

For the second sequence, I actually used the decimal expansion of π to determine the toss:

$$3.1415926535897932384\ldots$$

If the digits were in the region 1 to 5, I chose heads; otherwise I chose tails. The decimal expansion of π is conjectured to be random. It certainly has no discernible patterns. It has been proved to be an irrational number, which means that at no point does the expansion settle into a repeated sequence of digits, unlike $1/7=0.142857142857\ldots$ It is thought to be an example of a 'normal number', which means that each digit appears on average 1/10th of the time but that still hasn't been proved either way.

At first sight, the decimal expansion of π seems to be favouring small digits since heads cropped up many more times, but this is one of the dangers of small data sets. If you look at the first 32 digits of π, then the number 0 doesn't appear at all. And yet, in the long run, we still expect 0 to appear 1/10th of the time.

There are many tests that have been devised to test sequences for randomness, and π passes them all. These tests are very important in the modern age because many security protocols on the internet require a computer to be able to generate random sequences of numbers that no one can guess. Computers face an interesting dilemma. Cage could use the physical unpredictability of the toss of a coin to generate his random sequence. A computer has to depend on the programmer to write a bit of code to produce randomness. But that means there is a formula that is responsible, which means it isn't really random.

There is a concept of 'pseudo-randomness', which is a process that, although outwardly random, is in fact deterministic and repeatable.

The output passes all the tests one might throw at the sequence for randomness, and yet anyone with complete knowledge of the system can determine the results. Physical processes under classical Newtonian physics should actually be regarded as pseudo-random. As Laplace indicated, if you know the starting conditions of the toss of a coin then the equations of motion will predict the outcome. The point is that many of these physical processes are chaotic and so tiny perturbations in the initial conditions can result in wildly varied outcomes. Computer programs that generate random numbers exploit the same principle of using non-linear, chaotic equations as the source of the output.

Cryptographic systems often require random numbers to ensure that patterns don't emerge that make messages vulnerable to attack. Increasingly, lotteries are using computers to generate the winning numbers. In the past, a tumbling cage of numbered balls would be used, which exploited physical randomness, but today the numbers are generated by the click of a button. But care needs to be exercised. In 2005 Eddie Tipton slipped some malicious code into the software he had access to at the Multi-State Lottery Association in America, which diverted the code used to pick random numbers based on readings from a Geiger counter and instead inserted numbers that he could predict. The fraud was eventually picked up when a shadowy figure that matched Tipton's description was caught on camera buying the winning ticket.

A Geiger counter is potentially the only truly random way to generate unpredictable numbers. Modern physics asserts that the quantum world is the only place where we see genuine randomness at play. If I take a piece of uranium, then the physics says there is no mechanism to determine when the uranium will spit out its next bit of radiation. Whether this is truly random has been a matter of debate ever since the theory transformed physics at the beginning of the twentieth century. Einstein famously hated the idea that physics embraced randomness as part of its make-up. 'The theory produces a good deal but hardly brings us closer to the secret of the Old One,' he wrote in December 1926. 'I am at all events convinced that *He* does not play dice.'

One of the interesting concepts that has been applied to test randomness is that of information. Our first sequence of alternating heads and tails doesn't contain much information. Essentially the message is H T repeated ten times. The second sequence contains much more information. The measure of the information contained in a message is whether the message can be replicated by an algorithm that is shorter than the original message. This compression of the data is a sign that there is an underlying pattern that betrays the non-random nature of the sequence.

This might suggest an unconscious motivation within modern artists to find randomness attractive. The artistic patterns and structures that have been used before actually are limiting the information that can be expressed. Randomness is opening the artist up to new messages. It's like expanding the number of possible words that can appear in a dictionary. The new strings of letters that randomness might add give the possibility of new expression. The only challenge is that if you flood the dictionary with all possible strings then it becomes meaningless, much like the books in Borges's 'The Library of Babel'.

'Entropy' is a way to measure the information – or the uncertainty or surprise – in data. The idea of entropy was first introduced into physics to describe the movement from order to disorder that the universe experiences, as expressed in the second law of thermodynamics. The ordered structure of a sandcastle will decay over time into a disordered pile of sand. Entropy essentially is a measure of the possible states that all the sand can be in. When the sand is confined to the shape of a castle, the possibilities are limited. As a pile of unshaped sand, the possibilities are far greater. Entropy measures this. It was the American mathematician Claude Shannon who suggested in a paper published in 1948 that the concept of entropy could be applied to information as much as to physical matter.

The rather curious run of heads that begins the sequence of coin-tosses generated by π is often interpreted by the inexperienced onlooker as some sort of bias in the coin being used. But clustering of data is actually a common feature of truly random systems. Often the absence of clumping can be an indication that someone has been tinkering with the data.

Lotteries, for example, will often have results where there are two consecutive numbers on the winning ticket. The mathematics reveals that if you select six numbers at random from 49 numbers then half the possible choices have consecutive numbers. But if you ask people what numbers they choose when playing the lottery, then significantly less than half choose consecutive ones. We seem to resist this clumping, feeling that it is unlikely that numbers so close will come up together in the draw. This demonstrates how bad we are at behaving in a truly random fashion.

Are artists really random?

When my wife and I tiled our kitchen recently with a range of different coloured tiles, we wanted a random look. But when I used the decimal expansion of π to determine the sequence, my wife was not happy with the fact that every now and again there were three red tiles together. My initial arrangement was tampered with to destroy this clumping. It means that although the tiles in our kitchen might look random, they actually reflect the creative eye of my wife.

The Turner Prize-winning artist Mark Wallinger faced a similar challenge when he built a wall out of 10,000 individually numbered bricks. Rather than placing them in numerical order, he wanted to randomly arrange them. Why the piece resonates with people is that our immediate response is to search out patterns and intentions in the arrangement of the numbers. Our brains are so addicted to looking for patterns that we start to think there is a logic to the construction. For example, if we see two consecutive numbers lined up in the wall, then our first impression is that this must be planned. But my analysis revealed that there is an 86 per cent chance that any arrangement of these bricks will throw up a wall containing consecutive numbers. For Wallinger, the wall is a way of engaging with the concept of a very large something that humans have very little intuition for. It is often because we don't have experience of what one in a million really looks like that we find probability so counter-intuitive. Wallinger is intrigued by the way that 10,000 bricks when put together become a wall. 'There are 10,000 bricks, all individually numbered, but we call it a wall so we can deal with it. It's like an equation in mathematics.'

The German artist Gerhard Richter is also fascinated by this tension between generating something using randomness and the way it still provokes the viewer into searching for meaning. In his 2007 piece called *4900 Farben*, or *4900 Colours*, he created 196 square paintings, each consisting of 5×5 colour squares. He had a total of 25 colours to choose from, using a computer running a random number generator to pick which one was used in each square.

Richter has been experimenting with colour and randomness since he created his first grid paintings in 1966. In earlier versions of these, he selected which of his 25 colours to use in each square by pulling it out of a bag. He then returned the colour to the bag so that it would be possible for the same colour to be used again in the next square along. This analogue approach to the choices being made was matched by Richter individually painting each square. But in the most recent iteration from 2007 everything had been automated. The paint was now applied mechanically using a spray gun. That decision is a statement in itself of the impact of technology on the way society operates.

What do these paintings mean for Richter?

> They are the only paintings which tell no story. Even abstract paintings are like photos of a non-existent reality, of an unknown jungle. Here there is no illusion. They say nothing and evoke no association. They are simply there, pure visual subjects.

Not everyone is a fan. 'What arrogant tosh,' wrote Andrew Lambirth in the *Spectator*. 'Richter pontificates as if he's just invented geometric abstraction.' But the point is that Richter has been trying to produce art to deal with trauma in a post-Nazi Germany. Just as the Dada movement was a response to the horrors of the First World War, Richter is searching for a way to respond to what the art critic Benjamin H. D. Buchloh called the 'destruction of bourgeois subjectivity and ideology caused by World War II and the Holocaust'.

In an earlier exhibition of his colour paintings at the Tate, Richter quoted John Cage: 'I have nothing to say and I am saying it.' Randomness is the perfect vehicle for saying nothing. The curious thing is how we are so keen to see messages and meaning that we fill even randomness with our own insights.

For Richter, this is the charm of his colour paintings: 'What I like about the patterns are they're not constructed on the basis of an ideology or religion.' The trouble is that most ornamentation that artists use in their work already has meaning that originates from some other context. Richter felt that randomness was our best shot at breaking that connection. It was still inevitable that we would make visual connections in the randomness when we engaged with the work but those connections didn't originate in the artist's intention but rather the viewer's world of association. Richter believed that the patterns emerging from randomness 'are more universal. That's what I liked about them.'

I was very intrigued by whether Richter had truly given himself over to picking colours randomly or whether, like my wife choosing the tiles in my kitchen, he too couldn't resist some artistic tampering. At the same time as he was producing *4900 Colours*, Richter was also commissioned to create a huge stained-glass window for Cologne Cathedral. He used the same idea of random colour choices, but then added some symmetry to the window. Its left-hand side was chosen randomly, while the right-hand side is a reflection of it, making something rather like a stained-glass Rorschach inkblot.

But in the stained-glass window that the randomness produced there were regions that looked as if they were encoding messages. Richter decided to swap out two blue squares precisely because, alongside other squares, they were creating a very obvious image of the number 1. 'Once you've seen it, it is always there. One doesn't want a lightning bolt or a Swastika in the design either.' Perhaps that is understandable in the case of Cologne Cathedral. But what about in *4900 Colours*? Was there any similar tinkering to remove patterns that had appeared randomly?

What is the chance of generating a row of five squares with no consecutive colours? You can choose any colour for the first square. Then there is a 24/25 chance that the next colour doesn't match. Similarly, for the third square, it simply has to avoid the previous colour so that is again 24/25 chance. Over the five squares, we should have a $(24/25)^4$ chance of getting a row with no consecutive colours. Out of the 196×5 rows that means 85 per cent will not have consecutive squares of the same colour. But that means we should see 15 per cent, namely 147 rows, which do contain consecutive colours.

It turns out there are exactly 147 rows with consecutive squares of the same colour. Spooky. You'd expect to get a little error either side of this. The various other tests I ran on the paintings all supported the claim that the colours were chosen randomly.

Although Richter gave up any agency in the arrangement of the colours, there are some suggestions for how the paintings are displayed. He has 11 different versions that can be shown. Version I sees all 196 paintings hung separately. But in Version II, four paintings are put together to make a 10×10 grid of which there are 49 in total. Richter does not specify which paintings must go together, nor which way up they must be hung, which ends up creating a huge number of possible combinations. Instead the configuration is determined by Richter casting dice. Other versions have different dimensions in order to divide the 196 paintings into further combinations of squares. For example, Version X has three squares made up of 25 paintings, which leaves 121 paintings to be arranged in an 11×11 configuration. The final Version XI collects all 196 paintings into a single huge 14×14 array. The only choice a gallery has is which version of *4900 Colours* to display.

The mathematician in me couldn't help wondering how many other ways there are to arrange the paintings in addition to the 11 versions that Richter had chosen. Hiding behind them all is the question of how many different ways 196 can be divided up into sums of squares. Most of the versions don't display any single paintings by themselves. In addition to Version 1, where they are displayed individually, there is one more exception. Version V has:

$$196=5^2+5^2+5^2+5^2+5^2+5^2+5^2+4^2+2^2+1.$$

I thought that was rather unsatisfactory. So I decided to see how many ways there were to arrange the 196 canvases into square paintings whose dimensions were at least 2×2. After a mathematical analysis, I found that there are in fact 643 different ways. For example here are a couple of combinations that Richter missed:

$$196=12^2+6^2+4^2$$

or

$$196=13^2+3^2+3^2+3^2.$$

The question of which numbers can be written as sums of squares has an ancient heritage and has been an important theme throughout mathematical history. Fermat proved that any prime number that has remainder 1 when divided by 4 can be written as a sum of two squares. For example:

$$13=3\times4+1=2^2+3^2$$

or

$$41=4\times10+1=4^2+5^2.$$

This is one of my favourite theorems in mathematics. It has that beautiful narrative quality that I'm looking for in a good proof. What on earth do prime numbers and square numbers have to do with each other? And yet the mathematics takes you on a logical journey where one character, primes, gradually mutates throughout the story to show a completely different side of its character. A mathematical version of the *Strange Case of Dr Jekyll and Mr Hyde.* It was one of the proofs that I showed Emily Howard, when we created *Four Musical Proofs and a Conjecture*, to illustrate the idea of algebraic transformation.

Fermat announced the discovery to his friend Marin Mersenne on Christmas Day 1640. The proof used an equation discovered by the Indian mathematician Brahmagupta in the seventh century CE which shows that, if you take two numbers which can both be written as sums of two squares, then so can the product of these two numbers:

$$(a^2+b^2)(c^2+d^2)=(ac-bd)^2+(ad+bc)^2=(ac+bd)^2+(ad-bc)^2.$$

For example:

$$(1^2+4^2)(2^2+7^2)=30^2+1^2=26^2+15^2.$$

Fermat was not good at writing down his proofs. He famously wrote in his copy of Diophantus' *Arithmetica* that he had a remarkable proof of his so-called Last Theorem, but the margin was too small to contain it. When it came to his discovery about primes and squares it was the great eighteenth-century mathematician Leonhard Euler who was the first to flesh out the underlying story and prove Fermat's theorem. In 1770, the Italian-born French mathematician Joseph-Louis Lagrange proved that every number can be written as the sum of four square numbers. And provided the number is not of the form $4^k(8m+7)$, it can be written as the sum of three square numbers.

The question of how many different ways there are to write a number as a sum of an arbitrary number of squares as opposed to just three or four relates to important analytic functions called 'Jacobi theta functions'. In the case of Richter's 196 paintings I could calculate the answer to how many ways there are to gather them into squares by hand but it is fascinating to see how Richter's restrictions lead one naturally into deep questions about number theory. Once again, mathematics and art find themselves drawn to the exploration of common structures.

These restrictions – in terms both of the limited number of versions that can be displayed, and also of the rigid 5×5 squares that the colours are confined to – creates a strange tension with the randomness of the colour choices. 'I have nothing to say,' Richter had boldly declared with his colour paintings but perhaps this dialectic between chance and structure is at the heart of why this piece actually is not saying nothing.

Richter's artistic output is particularly striking for its range of different styles, from his photo paintings in which he paints over photographs to more abstract works like his *4900 Colours*. Since the 1970s, he has developed another very distinctive style of abstract painting which involves layering paint on the canvas using what he calls a squeegee. He starts by covering the canvas with a detailed underpainting in oils which is then obscured by the addition of more and more paint, applied randomly. The additions are then smeared and scraped across the underpainted image with a variety of different tools, including sponges and wooden blocks. But his favourite tool is the squeegee made from a long strip of flexible Perspex with a wooden handle. 'With a brush you have control. The paint goes on the brush and you make the mark . . . with the squeegee you lose control.' Another example of how Richter enjoys surrendering the creation of the piece to some external agent.

Many of his techniques are aimed at surprising himself with something new. As he says, 'I want to end up with a picture that I haven't planned.' That's why employing chance or using a brush which he can't control is going to produce something that was not predetermined. 'I just want to get something more interesting out of it than those things I can think out for myself.'

The abstract paintings using this squeegee technique are stunning and have some resonances with Pollock's work because of the extraordinary amount of complexity that is created by the process. Richter came up with a rather ingenious way to understand this complexity. It took its lead from the idea of mirroring the randomness in the stained-glass window in Cologne Cathedral. He chose one of his hugely complex abstract pieces from 1990 and divided the painting vertically into two halves, each of which he then mirrored to create two new paintings. The symmetry in each new work contrasted with the highly random and chaotic nature of the paint on the original canvas.

At this point the image was reminiscent of the idea of a Rorschach inkblot. But then Richter decided to repeat the process. This time the original image was divided into four vertical strips. Each one was taken individually and repeated four times within the outline of the painting, reflecting the image each time as it went. The effect is to create a symmetrical image repeated twice now. It's symmetry within symmetry.

Now that the algorithm was up and running, it was clear where Richter would take it next. Divide the painting into eight vertical strips. And use each strip to create a new image where the strip is repeated eight times, again reflected each time it is laid down. The process has generated eight new works that are a mix of symmetry and chaos.

The striking thing is that once you push this algorithm to its limit, dividing the image vertically by two each time, there comes a point where the chaos in the horizontal direction disappears. Instead, by the time you've halved the image 12 times, creating 4,096 different vertical strips, and then taken each of these strips individually and repeated it across the canvas, reflecting the piece each time, you are left with an image that is a sequence of horizontal stripes. A vertical stripe that is 1/4096th of the whole image has reached a point where the horizontal structure at any point is a single colour. The repetition across the work then simply smears this colour horizontally across the image to create the coloured stripes.

The complexity of the structure of the original painting which formed the basis of the whole process is exposed by the need to go down to this level of division before the horizontal structure disappears. Perform the same process with a less complex starting image and the same stripes would appear much earlier. But what is rather beautiful is the way that the process applied to Richter's original painting creates images with simplicity in the horizontal direction but retains the complexity in the vertical direction. The result captures both order and randomness in one canvas. The horizontal complexity of the original painting is hidden in the fact that if you take any of the 4,096 different stripe paintings that have been created, they are all different. The colour palette varies hugely from one to the other. The complexity of the original has created 4,096 different images. Queneau would have approved of this algorithm for making one work the source of many.

Faking agency in AI

Richter used a computer to make his choices with *4900 Colours*. So should the computer in fact be credited as the artist here, or at least be given a share of the credit? The role of computers in creating art has gone through something of an explosion in recent years with the advent of machine learning and generative AI. But even before this recent revolution there have been attempts to create an algorithmic artist.

In order to try to create the illusion of agency in the computer, programmers will often resort to using some element of randomness or chance in the algorithms. The British-born artist Harold Cohen created an algorithmic artist that he called AARON and would often talk about the 'decisions' that AARON was making, as if the computer had agency. The code that ran AARON was written in a top-down manner so that Cohen wrote the instructions which the machine then carried out, so at first sight it looked like Cohen was in complete control of these decisions. But on further investigation of the code, one discovers that a random number generator was used at various points in the machine's decision-making.

The randomness was used by Cohen to create some distance between him and the machine. His inability to predict its actions gave the illusion that the machine was deciding them. But this was just a trick. The randomness removed agency from both entities: machine and human.

One of the first people to contemplate the role of machines in artistic creativity was the Victorian mathematician Ada Lovelace. Her mother was very keen to expose her to the scientific innovations of the time and took her to see Charles Babbage's Analytical Engine in action. This was a machine that Babbage had intended to speed up tedious mathematical calculations. But when Lovelace saw it, she began to speculate about its potential to do so much more than just long division to lots of decimal places. 'The engine might compose elaborate and scientific pieces of music of any degree of complexity or extent.' However, she also offered a word of caution about its role in any act of creativity:

> It is desirable to guard against the possibility of exaggerated ideas that might arise as to the powers of the Analytical Engine. The Analytical Machine has no pretensions whatever to originate anything. It can do whatever we know how to order it to perform.

Recent advances in the way code is written have now raised the possibility that it could in fact create something that we had not ordered it to perform. Machine learning is a process where code changes and mutates as it interacts with more data. Out of that learning process might arise new code that has some autonomy from the humans who wrote the original code. That has inspired the idea of a new challenge to test AI creativity, similar to the idea of the Turing test for machine intelligence.

The Lovelace test asks whether an algorithm can originate a creative concept or work of art in such a way that the programmer or anyone who has access to the program is unable to explain how the algorithm produced it. But there is an important addendum to the Lovelace test: the outcome must be repeatable. This clause is specifically there to exclude the use of chance as a way of ensuring that humans can't explain the output. The algorithm can't use a random number generator, or the vagaries of the weather or the radiation from a piece of uranium to determine the outcome. The creativity must come from the code itself. To pass the Lovelace test, the blueprint for creativity must be deterministic not dependent on randomness.

And yet as we have seen, we humans do enjoy giving up our agency to the randomness of a dice roll or the chance flip of a coin. It is a blueprint for breaking our dependence on structures that have become ingrained in our creative process. It has proved a helpful way of stopping ourselves from getting stuck in mechanistic behaviours. We can sometimes end up behaving more like machines than creative human beings. Just as Jean Arp and the founders of Dada had hoped, tapping into chance in the creative process is a powerful blueprint for pushing us into new unexplored territories. Chuck everything in the air and see what new blueprints emerge.

Finale

Mathematics. Nature. Art. The triangle that forms the blueprint for the thesis of my book. At one corner is mathematics, for me defined as the study of structure. This takes us to the second corner, where we find nature: a universe built from the structures of mathematics. As we move to the third corner of the triangle, we find artists responding to the universe around us, embedding, reinventing, exploiting the structures of mathematics and nature in their creative output. But that creative mindset in turn fuels the mathematics that we discover at the first corner.

Underpinning each of these three realms are the blueprints that we have encountered throughout the book. Each corner provides a different perspective on these fundamental structures. The Fibonacci numbers connected to nature's spirals and to Le Corbusier's Unité. Prime numbers to cicadas in North America and to Messiaen's *Quartet for the End of Time*. The geometry of the torus to the shape of the universe and to Bach's Goldberg Variations. Fractals to the growth of trees and to Pollock's paintings. The icosahedron to the structure of a virus and to Laban's dance scales.

But I wonder whether the two corners of mathematics and art might have an edge over nature. One of the things I love about mathematics is the ability to go beyond what nature allows. The abstract character of mathematics means it can live in the mind without having to have a physical presence. For the artist too, a story doesn't have to be real, a painting doesn't have to depict reality, music doesn't have to be a soundtrack for a journey through the world. I think this freedom to make other worlds that are self-consistent yet independent of our reality is perhaps something that sets mathematics apart from the other sciences and again connects it more with the creative arts. That might seem strange given that mathematics is bound by the strict rules of logic and there is no flexibility to make something true if a proof clearly indicates it is false. And yet, within those constraints, there is a huge amount of leeway to roam and create.

As Stravinsky recognised:

> My freedom consists in my moving about within the narrow frame that I have assigned myself for each one of my undertakings. I shall go even further: my freedom will be so much the greater and more meaningful the more narrowly I limit my field of action and the more I surround myself with obstacles.

The same happens in mathematics. The constraint that an even number can't be odd is crucial to our discovery of irrational numbers, for example.

The only limits on that creativity in mathematics are the concept of truth and the mathematician's imagination. This, for me, contrasts with the sciences, where there is an extra limiting factor: the theory must match the reality of the world we live in. If an experiment contradicts the theory, then the theory is thrown out, even if it is a beautifully consistent theory. If it doesn't match the world that science is trying to describe, then it is rejected, however elegant it might be. Its mathematical truth is not enough.

The physicist Paul Dirac summed up this tension between mathematicians free to create many worlds and scientists bound to understand just one world:

> The mathematician plays a game in which he himself invents the rules while the physicist plays a game in which the rules are provided by nature, but as time goes on it becomes increasingly

evident that the rules which the mathematician finds interesting are the same as those which nature has chosen.

That second point is very striking: the way that mathematical flights of fancy can actually lead us to new scientific insights into the physical space we inhabit. Dirac's own work is a perfect example of this. He was very much a mathematical physicist at heart and his mathematical imagination led him to discover surprising things about the physical universe. For example, an equation like $x^2=4$ actually has two solutions, $x=2$ and $x=-2$. Dirac understood that these quadratic equations describe matter. The positive solution was the matter we saw around us. But he proposed that the negative solution might correspond to a new sort of matter. At first, most rejected the idea as crazy. It turned out, though, that Dirac was right. His proposal led to the discovery of antimatter.

For Dirac, the beauty of equations was paramount. 'It is more important to have beauty in one's equation than to have them fit experiment.' You can really hear the mathematician speaking. Beauty certainly has been a powerful guide in navigating the truths underlying nature. The strange menagerie of fundamental particles from which the universe is made was unified into some sort of coherent order by mathematicians' understanding of incredible symmetrical objects in high dimensions. The way the particles manifested themselves as facets of an exotic hyper-dimensional shape led to the discovery of new particles that completed the symmetry.

Although beauty has been a powerful guide to help us understand the universe, there is a real tension in modern physics at the moment about whether everything can so easily succumb to beautiful equations. The quest for a theory of everything that neatly describes the universe may not exist. There is a feeling that maybe physics is actually more like biology than mathematics. Perhaps the properties of the fundamental particles can't be neatly explained but just rather randomly assumed the values they have in much the same way that certain animals like cats evolved rather than unicorns. Although there is some convergence built into evolution, which is why we see eyes appearing independently in 50 different species, much of the evolutionary tree is very random. There really is no reason why we couldn't have had many of the beasts of mythology, but no biologist is going to waste time describing the evolution of animals that don't exist. Random mutation took the animal kingdom in one direction rather than another. Physics might be the same.

But even if that is true, everything we see around us is still an exploration of what is possible within the bounds of certain structures that define the universe's constraints. That is why even the wild and untamed trajectory that evolution has taken is just one possible path within this structural framework.

For me, mathematics is the realm of the possible, nature the realm of the actual. The mathematical structures we have explored have an existence outside the physical. The natural world we see around us is a physicalised version of these abstract structures. The universe is a bit like the Alhambra. Each particular decorative pattern on the wall is a choice of how to realise the symmetrical blueprint offered by mathematics. The physical universe is just one instantiation of the mathematical universe, a decision of how to decorate the walls according to the blueprints available.

Perhaps I have been rather down on nature as poor cousin to the mathematical and artistic exploration of these eternal blueprints. It's true that nature provides an inferior approximation of a circle, fails to be truly fractal and is stuck in a finite world that can't count to infinity. And yet without nature's physicalisation of these blueprints – even if it is imperfect – would the mathematicians or the artists have come up with them independently of their physical realisation?

Our mathematics has emerged out of our attempts to navigate and predict and tame and manipulate the world we live in. The mathematics of ancient Egypt grew out of the need to understand the patterns of the flooding of the Nile, to calculate areas of land to tax them, to count the number of stone blocks needed to raise a pyramid. It led to the discovery of π and the square root of 2.

But then we started to enjoy playing with these mathematical structures in their own right. That playful attitude inspired the discovery of imaginary numbers, symmetrical shapes in hyperspace, the proof that Fermat's equations have no solutions, geometries that seem to describe non-existent worlds. But the truth is that all these imaginary structures emerged out of first experiencing the physical structures of nature. Imaginary numbers grew out of exploring square roots beyond those that measure across a square. Symmetrical shapes in 11 dimensions are an evolution of playing with the three-dimensional symmetries that crystals form.

What is perhaps so surprising is seeing these imaginary shapes and numbers becoming fundamental tools to help us explain the universe in which we live. Negative numbers led us to antimatter. Imaginary numbers are the key to the quantum world. Non-Euclidean geometries turn out to describe the fabric of space-time. Higher-dimensional symmetries explain the fundamental particles that exist.

The challenge is whether there are structures in the abstract Platonic realm that we as embodied human mathematicians can never access because they are beyond any connection to the physicalised universe. Maybe, because of our embodiment, human mathematicians can access only part of the universe of structure. Are there new blueprints out there waiting for us to uncover? To discover them, what is certainly true is that we will need to tap into the creative artist that hides inside us all.

Bibliography

The bibliography contains much of the material that helped inform the writing of this book. The following are some more specific references to source material.

Overture

The Semir Zeki quote is from a press release from UCL to accompany the publication of the paper by Zeki, Romaya, Benincasa, Atiyah (2014). The Philip Glass quote is from David Cunningham's essay 'Einstein on the Beach' published in Kostelanetz (1997).

Blueprint One

The story of the first performance of the *Quartet for the End of Time* has been documented in many interviews with the performers and members of the audience. These can be found in Hill (2008), Rischin (2003), Pople (1998). Several good papers on acoustics including Cox and d'Antonio (2003) provided material for the role of primes in building concert halls. The source for the quote by James Garnon about the acoustics of the Globe is from an article in 2017 by Matt Trueman on 'Building Shakespeare's Globe' on the Globe Theatre website. There is an extensive literature on the wonderful prime number cicadas including Karban, Black and Weinbaum (2000). Jack Dee's biography is sourced from the MacTutor website at the University of St Andrews that has a comprehensive list of biographies of mathematicians. As well as conversations on the football pitch and at lunch in New College, the section on Shakespeare was informed by reading various articles including Desai (2015). David Bennett's videos on YouTube on his channel David Bennett Piano were a helpful source for songs that use interesting rhythms as were Joe Shadid's videos on Reverb. With reference to *Mission Impossible*, Lalo Schifrin told the *Spokesman-Review* in 1996 'the only people who could dance to it would be people who have five legs or mutants who have one leg shorter than the other'. The SETI Institute website has a description of the different elements that make the Arecibo message as does the paper published by the staff at the National Astronomy and Ionosphere Center (1975).

Blueprint Two

Several good biographies together with Borges's own writing and interviews informed the passages on Borges's life including Wilson (2006) and Williamson (2004). The story of Protogenes and Apelles was first told by Pliny and can be found retold for example in Guillaume Apollinaire's article 'On the Subject of Modern Poetry' (1912). A conversation over dinner at New College with the musicologist Jeremy Summerly alerted me to the paper 'Trachtenberg' (2001) which describes the connection between Du Fay's piece *Nuper Rosarum Flores* and the dimensions of the cathedral in Florence. The modern theory of the construction of the Duomo in Florence can be found in Paris, Pizzigoni, Adriaenssens (2020). Boullée (1976) describes the ideas behind the cenotaph to Newton. Architecture Daily also has a good article about Boulée's cenotaph published on 10 September 2014. The story of the Goldberg Variations is from an early biography of Bach by Johann Nikolaus Forkel. Blacklock (2020) provided helpful information on Hinton's biography. The imdb website has details of the film *Moebius*.

Blueprint Three

Le Corbusier detailed much of his own story in his writing including the two volumes *The Modulor*, Le Corbusier (1948) and *Modulor II*, Le Corbusier (1955). Other material is sourced from biographies of Le Corbusier including Weber (2008), Curtis (1986) and Sołtan (1987). The quote by Jean-Marc Drut is from Frearson (2014). Information about Indian rhythms was sourced from Messiaen (1994), Singh (1985), Deo (2007). The quote from the Sanskrit play *Shakuntala* is taken from Anish Shah's contribution to the Wikipedia page on Arya Metre. The 16 metre line of verse by Hemachandra is from Deo (2007). The Darwin story about the elephants and the Tribonacci numbers is described in Podani, Kun and Szilágyi (2018). Philip Glass has written and talked about his music extensively including Glass (2015), Kostelanetz (1997) and Duckworth (1995). Other material comes from a panel session I did with Glass at the New York Science Festival.

Blueprint Four

Catton's account of the writing of *The Luminaries* is recorded in an article she wrote for the *Guardian* in April 2014. The story of Jacob Bernoulli is from the biographies of mathematicians at the MacTutor website at the University of St Andrews. Douady and Couder (1992) contains the research that connects flower growth to the golden ratio. The quote from Bejan about the golden ratio is from his paper 'Bejan' (2009). The Le Corbusier material is sourced from his own writing in *The Modulor* (1948) and *Modulor II* (1955). The material about the golden ratio in music is sourced from Howat (1983) and Lendvai (2000). Information about Mozart's connections to the masons that informed my event at the Royal Opera House was sourced from Chailley (1971).

Blueprint Five

Richard Taylor's work on the fractal character of Jackson Pollock's paintings can be found in Taylor, Guzman, Martin, Hall, Micolich, Jonas, Scannell, Fairbanks, and Marlow (2007), Taylor, Micolich and Jonas (1999), Taylor (2002), Taylor, Micolich and Jonas (2002) and Abbott (2006). There is also an article that Taylor wrote for the *Oregon Quarterly* on 1 January 2010 which documents his involvement in the 32 fake Pollocks. Examples of fractals in architecture were sourced from Shishin and Ismail (2016), Abdelsalam and Ibrahim (2019) and Patuano and Lima (2021). Loren Carpenter's story of the making of *Vol Libre* is from an interview I did with him when I visited Pixar's studios. The website mandelbulb.com has material about the use of fractals in the Marvel movies. The websites bugman123.com and skytopia.com have information about Paul Nylander and Daniel White's work on 3D fractals. The quote by Kevin Smith is taken from an article by VFX journalist Ian Failes from 2018 that can be found at https://www.creativebloq.com/features/guardians-of-the-galaxy-vol-2-behind-the-scenes. The use of fractals in the movie *Doctor Strange* is sourced from Steven Ornes's article in ScienceNewsExplores published on 9 January 2020 entitled: How math makes movies like Doctor Strange so otherworldy https://www.snexplores.org/article/math-movies-doctor-strange-otherworldly. Examples of fractals in music were informed by McDonough and Herczyński (2023). The quote from Tom Stoppard is from a public conversation with Robert Osserman at the Mathematical Sciences Research Institute in Berkeley. The David Foster Wallace quote is from a 1996 Bookworm interview with Michael Silverblatt.

Blueprint Six

Laban's biography was sourced from Preston-Dunlop (1998), McCaw (2011), Kew (1999), Manning (1988), Dickson (2016), Dörr and Lantz (2003), Brooks (1993), Karina, Kant and Steinberg (2004) together with his own extensive writing including Laban (1920), (1936). Emmer

(1982) and Kemp (1990) contain examples of the platonic solids and the use of mathematics in Renaissance art. The story of the rediscovery of the Archimedean solids is described in Field (1997) and Schreiber, Fischer and Sternath (2008). The sculptor George Hart has a good collection of images of the platonic and archimedean solids drawn by a range of Renaissance artists: georgehart.com/. Biographic details about Salvador Dalí were sourced from Dalí (1936), (1958), (1964), (1992), Banchoff (2014), Wildgen (2016), Patterson (2020) and A National Gallery of Victoria Educational Resource on Dalí. The quote by Dalí about *The Sacrament of the Last Supper* is taken from the National Gallery of Art website nga.gov. Alex Garland's account about the writing of *The Tesseract* is taken from an interview he did in June 1999 on KCRW, a National Public Radio member station broadcasting from the campus of Santa Monica College in Santa Monica, California. Information about Zvi Hecker's architecture is sourced from Hecker (1977) and (1980). Graves (1882) contains details of Hamilton's life including the story of the Isosian Game.

Blueprint Seven

The material about Raymond Queneau, Oulipo and the sestina is sourced from the following publications Andrews (2004), Fraser (2004), Bamford (2020), Queneau (1947), Roubaud (2007), Saclolo (2011), Aubin (1997), Ferraro (2011), Queneau (2002), Dumas (2008), Champneys, Hjorth, and Man (2018), Champneys (2018), Asveld (2013), Guicharnaud (1951), Motte (2006), Stump (1999). Xenakis's biography is sourced from Xenakis (1990), (2008), Varga (1996), Matossian (2005). The quote from Messiaen is from an interview that Matossian did for her biography of Xenakis. The mathematics of Messiaen's 'Ile de Feu II' is from Berry (1976) and Benson (2007). The *Marie Claire* quote about the performance of Messiaen's *Visions de l'Amen* is from Simeone (2000). Material that informed the section about Calvino's work include Moreno-Viqueira (2013), Fernandes and Silva (2014), Martins (2018).

section on B. S. Johnson. Meyer-Eppler's definition of aleatoric music is from Meyer-Eppler (1957). The story of the mistranslation is sourced from Jacobs (1966). Material and quotes about John Cage's work are sourced from Kostelanetz (2003), Duckworth (1995), Ryan (2009), Cage (1961). The analysis of Boulez's piece *Constellation-Miroir* is informed by Trenkamp (1976) and Boulez (1971). The analysis of Stockhausen's *Klavierstück XI* is informed by Truelove (1998). Hedges (1978) provided source material for the eighteenth-century examples of music composed by dice. The section on Aleatoric Architecture is informed by Keller and Jaeger (2016). The Barbican website barbican.org.uk/ has an article by Jon Astbury about the architecture of the Barbican. Eddie Tipton's story can be found in Reid Forgrave's article 'The Man who Cracked the Lottery' in the *New York Times* on 3 May 2018. The quote by Mark Wallinger is from an interview I did with him. The description of the process behind Richter's *4900 Colours* can be found in Richter (2007, 2008, 2009). The Richter quote 'What I like about the patterns . . .' is translated from an interview filmed by Anders Kold at Richter's studio in Cologne, Germany, 2016. Other quotes are from Richter (2007, 2008, 2009) and Richter's website gerhard-richter.com/.

References

Abbott, A. (2006) 'In the hands of a master'. *Nature* 439, pp. 648–50.

Abdelsalam, M. and Ibrahim, M. (2019) 'Fractal Dimension of Islamic Architecture: The case of the Mameluke Madrasas: Al-Sultan Hassan Madrasa'. *Gazi University Journal of Science* 32, pp. 27–37.

Abdullahi, Y. and Embi, M. R. B. (2013) 'Evolution of Islamic geometric patterns'. *Frontiers of Architectural Research* 2, pp. 243–251.

Andrews, C. (2004) 'Numerology and Mathematics in the Writing of Raymond Queneau'. *Forum for Modern Language Studies* 40, pp. 291–300.

Antokoletz, E. (1984) *The Music of Béla Bartók*. University of California Press.

Asveld, P. R. J. (2013) Queneau Numbers – Recent Results and a Bibliography. (CTIT Technical Report Series; No. TR-CTIT-13–16). Centre for Telematics and Information Technology (CTIT).

Aubin D. (1997) 'The Withering Immortality of Nicolas Bourbaki: A Cultural Connector at the Confluence of Mathematics, Structuralism, and the Oulipo in France'. *Science in Context* 10, 2, pp. 297–342.

Bamford, A. (2020) 'Mathematics and Modern Literature (Passages from "Chalk and the Architrave")' *New Left Review* 124, pp. 107–23.

Banchoff, T. F. (2014) Dalí and the Fourth Dimension. Proceedings of Bridges 2014: Mathematics, Music, Art, Architecture, Culture.

Barnatt J, Moir G. (1984) 'Stone Circles and Megalithic Mathematics'. *Proceedings of the Prehistoric Society* 50(1):197–216.

Bartlett, C. (2019) 'Nautilus Spirals and the Meta-Golden Ratio Chi'. *Nexus Network Journal* 21, pp. 641–56.

Bejan, A. (2009) 'The golden ratio predicted: Vision, cognition and locomotion as a single design in nature'. *Int. J. of Design & Nature and Ecodynamics.* Vol. 4, No. 2, pp. 97–104.

Benson, D. (2007) *Music: A Mathematical Offering.* CUP.

Bernard, J. (2003) 'Zones of Impingement: Two Movements from Bartók's Music for Strings, Percussion, and Celesta'. *Music Theory Spectrum* 25, pp. 3–34.

Berry, W. (1976) *Structural Function in Music.* Prentice-Hill.

Blacklock, M. (2020) 'The four dimensional life of mathematician Charles Howard Hinton'. BBC Focus Magazine.

Borges, J. L. (1970) *Labyrinths.* Penguin Books.

Borges, J. L. and Ferrari, O. (1986) *En Diálogo 1.* Editorial Sudamericana.

Borges, J. L. (1999) *The Total Library: Non-Fiction 1922–1986.* Allen Lane.

Boulez, P. (1971) *Boulez on Music Today.* Faber & Faber.

Boullée, E.-L., (1976) *Architecture, Essay on Art.* Academy Editions.

Broad, S. (2005) 'Recontextualising Messiaen's Early Career'. DPhil thesis, Oxford University.

Brookes, A. (2013) 'Non-Euclidean Geometry and Russian Literature: A Study of Fictional Truth and Ontology in Fyodor Dostoevsky's *Brothers Karamazov*, Vladimir Nabokov's *The Gift*, and Daniil Kharms's *Incidences*'. PhD, thesis Yale University.

Brooks, L. M. (1993) 'Harmony in Space: A Perspective on the Work of Rudolf Laban'. *The Journal of Aesthetic Education*, Summer, 1993, Vol. 27, No. 2, pp. 29–41.

Burroughs, W. (1963) 'The Cut Up Method' from Leroi Jones, ed., *The Moderns: An Anthology of New Writing in America*. NY: Corinth Books.

Burroughs, William S. (1986) 'Creative Reading', in *The Adding Machine*. New York: Seaver, pp. 37–46.

Burry, J. and Burry, M. (2010) *The New Mathematics of Architecture*. Thames and Hudson.

Cage, J. (1961) *Silence: Lectures and Writings*. Wesleyan Univ Press.

Calvino, I. (1969) *The Castle of Crossed Destinies*. Vintage.

Calvino, I. (1972) *Invisible Cities*. Vintage.

Calvino, I. (1979) *If on a Winter's Night a Traveller*. Vintage.

Calvino, I. (1983) *Mr Palomar*. Vintage.

Calvino, I. (2010) *The Complete Cosmicomics*. Penguin Modern Classics.

Calvino, I. (2023) *The Written World and the Unwritten World*. Penguin Modern Classics.

Carpo, M. (2011) *The Alphabet and the Algorithm*. The MIT Press.

Catton, E. (2013) *The Luminaries*. Victoria University Press.

Chailley, J. (1971) *The Magic Flute, Masonic Opera*. New York: A Knopf.

Champneys, A. (2018) 'Westward Ho! Musing on Mathematics and Mechanics'. *Mathematics Today*, October 2018, pp. 245–8.

Champneys, A., Hjorth, P. G. and Man, H. (2018). The numbers lead a dance: Mathematics of the Sestina. In Non-linear partial differential equations, mathematical physics, and stochastic analysis: the Helge Holden anniversary volume (EMS series of Congress reports). European Mathematical Society. pp. 55–71.

Christensen, T. (1993) *Rameau and Musical Thought in the Enlightenment*. CUP.

Coe, J. (2004) *Like a Fiery Elephant: The Story of B. S. Johnson*. Basingstoke and Oxford: Picador.

Costa, M. (2008) 'Woolly Thinking'. *Guardian*, 4 June 2008.

Cox, T. J. and d'Antonio, P. (2003) 'Engineering art: the science of concert hall acoustics'. *Interdisciplinary Science Reviews*, Vol. 28, No. 2, pp. 119–29.

Coxeter, H. S. M. (1979) 'The Non-Euclidean Symmetry of Escher's Picture "Circle Limit III"'. *Leonardo*, Vol. 12, No. 1, pp. 19–25.

Curtis, W. J. R. (1986) *Le Corbusier: ideas and forms*. Phaidon.

Dalí, S. (1936) *Conquest of the Irrational*. New York Julien Levy.

Dalí, S. (1958) *Anti-matter Manifesto*. Carstairs Gallery.

Dalí, S. (1964) *Diary of a Genius*. Doubleday.

Dalí, S (1992) *50 Secrets of Magic Craftsmanship*. Dover Publications.

Deo, A. S. (2007) 'The Metrical Organization of Classical Sanskrit Verse'. *Journal of Linguistics*, Vol. 43, No. 1, pp. 63–114.

Desai, A. N. (2015) 'Number-Lines: Diagramming Irrationality in "The Phoenix and Turtle"'. *Configurations*, Volume 23, Number 3, pp. 301–30.

Dickson, Christine (2016) 'Dance Under the Swastika: Rudolf von Laban's Influence on Nazi Power'. *International Journal of Undergraduate Research and Creative Activities*, Vol. 8, Article 7.

Doolittle, E. L., Gingras, B., Endres, D. M. and Tecumseh Fitch, W. (2014) 'Overtone-based pitch selection in hermit thrush song: Unexpected convergence with scale construction in human music'. PNAS, vol 11, no 46, pp, 16616–21.

Dörr, E. and Lantz, L. (2003) 'Rudolf von Laban: The "Founding Father" of Expressionist Dance'. *Dance Chronicle*, 2003, Vol. 26, No. 1, pp. 1–29.

Douady, S. and Couder, Y. (1992) 'Phyllotaxis as a physical self-organized growth process'. *Phys. Rev. Lett.* Vol 68, issue 13, pp. 2098–2101.

Drożdż, S., Oświęcimka, P., Kulig, A., Kwapień, J., Bazarnik, K., Grabska-Gradzińska, I., Rybicki, J., Stanuszek, M. (2016) 'Quantifying origin and character of long-range correlations in narrative texts'. *Information Sciences*, Volume 331, pp. 32–44.

du Sautoy, M. (2001) 'A nilpotent group and its elliptic curve: Non-uniformity of local zeta functions of groups'. *Isr. J. Math.* 126, pp. 269–88.

du Sautoy, M. (2002). 'Counting subgroups in nilpotent groups and points on elliptic curves'. *Journal für die reine und angewandte Mathematik*, (549), pp. 1–21.

Dubow, B. (2022) 'And That Misformed Shape, Misshaped More'. The Stranger Mathematics of The Faerie Queene. DPhil thesis, Oxford University.

Duckworth, W. (1995) *Talking Music: Conversations with Five Generations of American Experimental Composers*. Da Capo.

Dumas, J.-G. (2008) 'Caractérisation des quenines et leur représentation spirale'. *Mathématiques et sciences humaines*, 184, pp. 9–23.

Emmer, M. (1982) 'Art and Mathematics: The Platonic Solids'. *Leonardo*, Volume 15, Number 4, pp. 277–82.

Engelhardt, N. (2018) *Modernism, Fiction and Mathematics*. Edinburgh University Press.

Ernst, B. (1976) *The Magic Mirror of M. C. Escher*. Taschen.

Fernandes, E. and Silva, A. (2014) *From Moore to Calvino: The invisible cities of 20th century planning*. Escola Superior Artistica da Porto.

Ferraro, A. (2011). 'Writing and Mathematics in the Work of Raymond Queneau'. In: Bartocci, C., Betti, R., Guerraggio, A., Lucchetti, R. (eds) *Mathematical Lives*. Springer, Berlin, Heidelberg.

Field, J. V. (1997) 'Rediscovering the Archimedean Polyhedra: Piero della Francesca, Luca Pacioli, Leonardo da Vinci, Albrecht Dürer, Daniele Barbaro, and Johannes Kepler'. *Archive for History of Exact Sciences*, Vol. 50, No. ¾, pp. 241–89.

Forkel, J. N. (1802) *Johann Sebastian Bach: his life, art and work*. Leipzig: Hoffmeister und Kühnel.

Fraser, R. (2004) 'Past Lives of Knives: On Borges, Translation, and Sticking Old Texts'. *Traduction, terminologie, rédaction*, vol. 17, no. 1, pp. 55–80.

Frearson, A. (2014) 'Brutalist buildings: Unité d'Habitation, Marseille by Le Corbusier'. *Dezeen* magazine.

Fry, S. (2005) *The Ode Less Travelled*. Hutchinson.

Glass, P. (2015) *Words Without Music*. Faber & Faber.

Graves, R. P. (1882) *Life of Sir William Rowan Hamilton, Andrews Professor of Astronomy in the University of Dublin, and Royal Astronomer of Ireland; including Selections from his Poems, Correspondence, and Miscellaneous Writings*. Dublin University Press.

Guicharnaud, J. (1951) 'Raymond Queneau's Universe'. *Yale French Studies* 1951, No. 8, What's Novel in the Novel, pp. 38–47.

Guthrie, J. (2012) 'Laban – Space Harmony and Dance Movement Therapy'. *DTAA Journal*, Moving On, Volume 10, Nos 1 and 2, pp. 5–10.

Hadid, Z. (2014) 'Zaha Hadid RA on the influence of Malevich in her work'. *RA Magazine*, July 2014.

Hamouche, M. B. (2009). 'Can Chaos Theory Explain Complexity In Urban Fabric? Applications in Traditional Muslim Settlements'. In: Williams, K. (eds) *Nexus Network Journal*, vol 11, 2. Birkhäuser Basel.

Hecker, Z. (1977) 'The Geometry of My Polyhedral Sculpture'. *Leonardo*, Vol. 10, No. 3, pp. 183–7.

Hecker, Z. (1980) 'The Cube and the Dodecahedron in my Polyhedric Architecture'. *Leonardo*, Vol. 13, Number 4, pp. 272–5.

Hedges, S. A. (1978) 'Dice Music in the Eighteenth Century'. *Music & Letters*, Vol. 59, No. 2, pp. 180–7.

Henderson, D. and Taimina, D. (2001) 'Crocheting the Hyperbolic Plane'. *The Mathematical Intelligencer* 23, pp. 17–28.

Hieatt, A. Kent (1960) *Short Time's Endless Monument: The Symbolism of the Numbers in Edmund Spenser's Epithalamion, Short Time's Endless Monument* (New York: Columbia University Press).

Hill, P. (editor) (2008) *The Messiaen Companion*. Faber & Faber.

Hochgraf, K. A. (2019) 'The Art of Concert Hall Acoustics: Current Trends and Questions in Research and Design'. *Acoustics Today*, 2019, vol 15, issue 1, pp. 28–36.

Hornby, R. (2000) 'Shakespeare's Globe'. *The Hudson Review*,

Winter, 2000, Vol. 52, No. 4, The British Issue (Winter, 2000), pp. 633–40.

Howard, E. (2023) 'Torus – sphere – Antisphere Compassing Mathematical Curvature through Musical Composition'. *LMS Newsletter*, Issue 507, July 2023, pp. 23–7.

Howat, R. (1983) *Debussy in Proportion*. CUP.

Jackson, R. (2013). 'Dostoevsky's Concept of Reality and Its Representation in Art'. In *Close Encounters: Essays on Russian Literature* (pp. 239–60). Boston, USA: Academic Studies Press.

Jacobs, A. (1966) 'Admonitoric Note'. *The Musical Times*, Vol. 107, No. 1479, p. 414.

Janco, M. (1971) 'Dada at Two Speeds'. In Lucy R. Lippard, *Dadas on Art*. Englewood Cliffs, N. J.

Jenner, S. (2014). 'B. S. Johnson and the Aleatoric Novel'. In: Jordan, J., Ryle, M. (eds) *B. S. Johnson and Post-War Literature*. Palgrave Macmillan, London.

Jensen, M. G. (2009) 'John Cage, Chance Operations, and the Chaos Game: Cage and the "I Ching"'. *The Musical Times*, Summer, 2009, Vol. 150, No. 1907, pp. 97–102.

Jernigan, C. (1974) 'The Song of Nail and Uncle: Arnaut Daniel's Sestina "Lo ferm voler q'el cor m'intra"'. *Studies in Philology* 71(2), pp. 127–51.

Johnson, B. S. (1973) *Aren't You Rather Young to be Writing Your Memoirs?* London: Hutchinson.

Johnson, B. S. (1999). 'Introduction to Aren't You Rather Young to be Writing Your Memoirs?'. *Review of Contemporary Fiction* 19(3).

Kapoor, A. (2011) *Turning the World Upside Down in Kensington Gardens*. Koenig Books.

Karban, R., Black, C. A. and Weinbaum, S. A. (2000) 'How 17-year cicadas keep track of time'. *Ecology Letters* 3, pp. 253–6.

Karina, L., Kant, M. and Steinberg, J. (2004). *Hitler's Dancers: German Modern Dance and the Third Reich* (1st ed.). Berghahn Books.

Keller, S. and Jaeger, H. (2016). 'Aleatory Architectures'. *Granular Matter* 18.

Kemp, M. (1990) *The Science of Art: optical themes in western art from Brunelleschi to Seurat*. Yale.

Kew, C. (1999) 'From Wiemar Movement Choir to Nazi Community Dance: The Rise and Fall of Rudolf Laban's "Festkultur" Dance Research'. *The Journal of the Society for Dance Research*, Vol. 17, No. 2, pp. 73–96.

Kittelmann, U., Wertheim, M. and Wertheim, C. (2022) *Margaret and Christine Wertheim: Value and Transformation of Corals*. Wienand Verlag.

Kostelanetz, R. (editor) (1997) *Writings on Glass*. Schirmer Books.

Kostelanetz, R. (2003) *Conversing with Cage*. Routledge.

Laban, R.v. (1935) *Ein Leben für den Tanz*, Carl Reissner Verlag, Dresden (translated 1975 as 'A Life for Dance').

Laban, R.v., (1936) 'Meister und Werk in der Tanzkunst'. Deutsche Tanzzeitschrift.

Laban, R.v. (1920) *Die Welt des Tänzers: fünf Gedankenreigen*. W. Seifert.

Le Corbusier (1948) *The Modulor*. Faber & Faber.

Le Corbusier (1955) *Modulor 2*. Faber & Faber.

Lee, H. (2020) *Tom Stoppard: A Life*. Faber & Faber.

Lendvai, E. (2000) *Béla Bartók*. Kahn and Averill, London.

Linford, M. B., (1933) 'Fruit quality studies II. Eye number and eye weight'. *Pineapple Quarterly* 3, pp. 185–8.

Loach, J. (1998) 'Le Corbusier and the Creative Use of Mathematics'. *The British Journal for the History of Science*, Vol. 31, No. 2, Science and the Visual, pp. 185–215.

Manning, S. (1988) 'Reinterpreting Laban'. *Dance Chronicle* 11, no. 2, pp. 315–20.

Marinoff, L. (2017) 'Dada as Philosophical Practice and vice versa: reflections on the centenary of the Cabaret Voltaire'. In *New Frontiers in Philosophical Practice*. Cambridge Scholars Publishing.

Marmot, A. F. (1981) 'The Legacy of Le Corbusier and High-Rise Housing'. *Built Environment*, Vol. 7, No. 2, The Importance of Planning History, pp. 82–95.

Marsh-Soloway, M. A. (2016) 'The Mathematical Genius of F. M.

Dostoevsky: Imaginary Numbers, Statistics, Non-Euclidean Geometry, and Infinity'. PhD thesis University of Virginia.

Martins, A. I. C. (2018). 'Invisible cities: utopian spaces or imaginary places?'. *Archai*, no 22, pp. 123–52.

Matossian, N. (2005) *Xenakis*. Moufflon Publications.

McCaw, D. (editor) (2011) *The Laban Sourcebook*. Routledge.

McDonough, J. and Herczyński, A. (2023) 'Fractal patterns in music'. *Chaos, Solitons & Fractals*, Vol. 170, pp. 1–13.

Meer, A. (1990) 'Interview with Anish Kapoor'. *Bomb Magazine*, Winter 1990.

Melzer, A. H. (1973) 'The Dada Actor and Performance Theory'. *Art Forum*, Vol 12, no 4, December 1973, pp. 51–7.

Melzer, A. H. (1994) *Dada and Surrealist Performance*. Johns Hopkins University Press.

Messiaen, O. (1994) *Traité de Rythme, de Couleur, et d'Ornithologie*. Alphonse Leduc, Paris.

Meyer-Eppler, W. (1957) 'Statistic and Psychologic Problems of Sound', translated by Alexander Goehr. Die Reihe 1 ('Electronic Music'), pp. 55–61. Original German edition, 1955, as 'Statistische und psychologische Klangprobleme', Die Reihe 1 ('Elektronische Musik'), pp. 22–8.

Mitra, A. (1989) 'Origin and Development of Sanskrit Metrics'. The Asiatic Society, Calcutta.

Moreno-Viqueira, I. (2013) 'Invisible Mathematics in Italo Calvino's Le città invisibili'. PhD thesis, Columbia University.

Motherwell, R. (editor)(1981) 'The Dada Painters and Poets: an anthology'. Boston, Mass., G. K. Hall.

Motte, W. (2006) 'Raymond Queneau and the Early Oulipo'. *French Forum*, Vol. 31, No. 1, pp. 41–54.

Newlove, J. (1993) *Laban for Actors and Dancers: Putting Laban's Movement Theory Into Practice: a Step-by-step Guide*. Routledge.

Onderdonk, P. B., (1970) 'Pineapples and Fibonacci numbers'. *The Fibonacci Quarterly*, 8(5), pp. 507–8.

Paris, V., Pizzigoni, A., Adriaenssens, S. (2020) 'Statics of self-balancing masonry domes constructed with a cross-herringbone spiraling pattern'. *Engineering Structures*, Vol. 215, pp. 1–10.

Parsley, J. and Soriano, C. T. (2009) 'Understanding geometry in the dance studio', *Journal of Mathematics and the Arts*, 3: 1, pp. 11–18.

Patterson, L. (2020) 'The Calculated Madness of Salvador Dalí's Mathematical Life: An analysis'. *ARTpublika Magazine*, Vol. 15.

Patuano, A., Lima, M. F. (2021) 'The fractal dimension of Islamic and Persian four-folding gardens'. *Humanit Soc Sci Commun* 8, 86, pp. 1–14.

Perle, G. (1989) *The Operas of Alban Berg*, Volume One: Wozzeck. University of California Press.

Perle, G. (2001) *Style and Idea in the Lyric Suite of Alban Berg*. Pendragon Press.

Podani, J., Kun, Á. and Szilágyi, A. (2018) 'How Fast Does Darwin's Elephant Population Grow?'. *J Hist Biol*, 51, pp. 259–81.

Pople, A. (1998) *Messiaen: Quatuor Pour La Fin du Temps*. CUP.

Potter, K. (2000) *Four Musical Minimalists*. CUP.

Preston-Dunlop, V. (1998) *Rudolf Laban: An Extraordinary Life*. Dance Books.

Queneau, A.-I. (2002) *Album Raymond Queneau: Iconographie Commentee (Albums de la Pleiade)*. Gallimard.

Queneau, R. (1933) *Le Chiendent*. Librairie Gallimard.

Queneau, R. (1947) *Exercises in Style*. (translated in English 1958) Gallimard.

Rian, I. M. and Sassone, M. (2014) 'Fractal-Based Generative Design of Structural Trusses Using Iterated Function System'. *International Journal of Space Structures*, Vol. 29, No. 4, pp. 181–203.

Richter, G. (2007) *Zufall: the Cologne Cathedral Window, and 4900 Colours*. Verlag Kölner Dom.

Richter, G. (2008) *4900 Colours*. Serpentine Gallery, Hatje Cantz.

Richter, G. (2009) *Texte zu 4900 Farben*. Hatje Cantz Verlag.

Richter, H. (1948) 'Dada XYZ . . .' Published in *The Dada Painters and Poets: an anthology*. Boston, Mass., G. K. Hall, 1981.

Rischin, R. (2003) *For the End of Time: the Story of the Messiaen Quartet*. Cornell University Press.

Roberts, S. (2024) 'The Crochet Coral Reef Keeps Spawning, Hyperbolically'. *New York Times*, 15 January 2024.

Roubaud, J. (2007) 'Bourbaki and the Oulipo'. *Journal of Romance Studies*, Vol. 7, Number 3, pp. 123–32.

Rozhkovskaya, N. (2019). 'Mathematical Commentary on Le Corbusier's Modulor'. *Nexus Network Journal* 22, pp. 1–18.

Ruppel, A. M. (2017) *The Cambridge Introduction to Sanskrit*. CUP.

Ryan, D. (2009) *Notes for recording of John Cage: Music of Changes*, performed by Tania Chen. Knitted Records.

Saclolo, M. P. (2011) 'How a Medieval Troubadour Became a Mathematical Figure'. *Notices of the American Mathematical Society*, Vol. 58, Number 5, pp. 682–7.

Salazar Sutil, N. (2013) 'Rudolf Laban and Topology: a Videographic Analysis'. *Space and Culture*, 16 (2), pp. 173–93.

Schattschneider, D. (2004) *M. C. Escher: Visions of Symmetry*. Thames and Hudson.

Schattschneider, D. (2010). 'The mathematical side of M. C. Escher'. *Notices of the American Mathematical Society* 57, pp. 706–18.

Schreiber, P., Fischer, G. and Sternath, M. L. (2008) 'New light on the rediscovery of the Archimedean solids during the Renaissance'. *Archive for History of Exact Sciences*, Vol. 62, No. 4, pp. 457–67.

Schroeder, M. R. (1979) 'Binaural dissimilarity and optimum ceilings for concert halls: more lateral sound diffusion', *Journal of the Acoustical Society of America* 65, pp. 958–63.

Schwarz, K. R. (1996) Minimalists. Phaidon Press.

Shishin, M. Y. and Ismail, K. (2016). A method of compositional fractal analysis and its application in Islamic architectural ensembles. Math Edu 11 (5), pp. 1087–1100.

Silverman, J. H. (1986). 'The Arithmetic of Elliptic Curves'. *Graduate Texts in Mathematics*, Vol. 106, Springer-Verlag.

Simeone, N. (2000) 'Messiaen and the Concerts de la Pléiade: "A Kind of Clandestine Revenge Against the Occupation" *Music and Letters*, Vol. 81, Issue 4, pp. 551–84.

Šimundža, M. (1988) 'Messiaen's Rhythmical Organisation and Classical Indian Theory of Rhythm (II)'. *International Review of the Aesthetics and Sociology of Music*, Vol. 19, No. 1, pp. 53–73.

Singh, P. (1985) 'The So-called Fibonacci Numbers in Ancient and Medieval India'. *Historia Mathematica* 12, pp. 229–44.

Slatoff-Ponté, Z. (2015) *Yogavatarannam*. North Point Press.

Sołtan, J. (1987). 'Working with Le Corbusier'. In: *Le Corbusier: The Garland Essays*, H. Allen Brooks, ed., pp. 1–16. New York: Garland Publishing.

Staff at the National Astronomy and Ionosphere Center (1975) 'The Arecibo message of November, 1974', *Icarus*, Volume 26, Issue 4, 1975, pp. 462–6.

Storr, R. (2009) *Cage: Six Paintings by Gerhard Richter*. Tate Publishing.

Stump, J. (1999) 'Reading Through the Manuscript: The Case of Queneau's Le chiendent'. *Dalhousie French Studies*, Fall 1999, Vol. 48, pp. 61–73.

Sylvester, D. (2016) *Interviews with Francis Bacon: The Brutality of Fact*. Thames and Hudson Ltd.

Taylor, R. P., Micolich, A. P. and Jonas, D. (2002) 'The Construction of Jackson Pollock's Fractal Drip Paintings'. *Leonardo* 35 (2), pp. 203–7.

Taylor, R. P. (2002), 'Order in Pollock's chaos', *Scientific American*, Volume 287, Issue 6, pp. 116–21.

Taylor, R. P., Guzman, R., Martin, T. P., Hall, G. D. R., Micolich, A. P., Jonas, D., Scannell, B. C., Fairbanks, M. S. and Marlow, C. A. (2007) 'Authenticating paintings using fractal geometry'. *Pattern Recognition Letters,* Volume 28, Issue 6, pp. 695–702.

Taylor, R. P., Micolich, A. P. and Jonas, D. (1999) 'Fractal analysis of Pollock's drip paintings'. *Nature* 399, p. 422.

Terry, P. (ed.) (2019) *The Penguin Book of Oulipo*. Penguin.

Toribio Vazquez, J. L. (2018) 'Narrative Circularity in Modern European Literature'. PhD thesis, University of Kent.

Trachtenberg, M. (2001) 'Architecture and Music Reunited: A New Reading of Dufay's "Nuper Rosarum Flores" and the Cathedral of Florence'. *Renaissance Quarterly*, Vol. 54, No. 3, pp. 740–75.

Trenkamp, A. (1976) 'The Concept of "Alea" in Boulez's "Constellation-Miroir"'. *Music & Letters*, Vol. 57, No. 1, pp. 1–10.

Truelove, S. (1998) 'The Translation of Rhythm into Pitch in Stockhausen's Klavierstück XI'. *Perspectives of New Music*, Vol. 36, No. 1, pp. 189–220.

Tubbs, R., Jenkins, A. and Engelhardt, N. (eds) (2021) *The Palgrave Handbook of Literature and Mathematics*. Palgrave Macmillan.

Tutte, W. T. (1946) 'On Hamiltonian Circuits'. *Journal of the London Mathematical Society*, Volume 21, Issue 2, pp. 98–101.

Varga, B. A. (1996) *Conversations with Iannis Xenakis*. Faber & Faber.

Vazquez, J. L. T. (2022) *Circular Narratives in Modern European Literature*. Bloomsbury.

Weber, N. F. (2008) *Le Corbusier: a life*. Alfred A. Knopf.

Wertheim, M., Henderson, D. and Taimina, D. (2005) 'Crocheting the Hyperbolic Plane: An Interview with David Henderson and Daina Taimina. Into non-Euclidean space, with hook and yarn'. *Cabinet Magazine*, Issue 16.

Wildgen, W. (2016). 'Intellectual revolutions in philosophy and art: continua and catastrophes'. Lo Sguardo. Rivista di filosofia.

Williamson, E. (ed.) (2013) *The Cambridge Companion to Jorge Luis Borges*. CUP.

Williamson, E. (2004) *Borges: A Life*. Viking.

Wilson, J. (2006) *Jorge Luis Borges*. Reaktion Books.

Xenakis, I. (1990) *Formalized Music: thought and mathematics in composition*. Pendragon Press.

Xenakis, I. (Translated, compiled and presented by Kanach, S.) (2008) *Music and Architecture*. Pendragon Press.

Yorke, J. (2014) *Into the Woods*. Penguin.

Young, K. (2020) 'Ivan Karamazov's Euclidean Mind: the "Fact" of Human Suffering and Evil'. *The Polish Journal of Aesthetics* 56. pp. 49–62.

Zeki S., Romaya J. P., Benincasa D. M., Atiyah M. F. (2014) 'The experience of mathematical beauty and its neural correlates'. *Frontiers in Human Neuroscience*, Vol 8, pp. 1–12.

Animation of the mathematics behind Xenakis's *Nomos Alpha* created with Simon Russell. https://www.youtube.com/watch?v=fWlce-olQMGQ.

Animation of the mathematics behind Messiaen's the *Quartet for the End of Time* created with Simon Russell. https://vimeo.com/151931854.

Acknowledgements

There is something of a fractal quality to publishing a book. The process has many layers but as you zoom in, each layer is as important as the last.

A book starts with an idea that I share with my brilliant agent Anthony Topping at Greene and Heaton. That seed takes on life as we nurture the idea. This book benefited from an inspired moment when Antony suggested flipping the thing upside down. From that moment the structure for the book was in place.

Writing my books has only been possible thanks to the next layer: Fourth Estate. An incredible publishing blueprint, they have been there ever since I first visited them in the piano factory in Notting Hill when they took a punt on my first book *The Music of the Primes*. The people at Fourth Estate love their books and their authors and I feel in very safe hands as I bring this 8th book to press.

As you begin to zoom in on the journey of the book the next layer you meet are my editors. I am especially lucky to have an incredible editor in Louise Haines. She has a brilliant eye for ensuring the overall structural blueprint of a book is working to tell the story I want to share. Louise is perfectly balanced by my editor across the pond, TJ Kelleher at Basic Books, who loves his mathematics and always encourages more "pencil up" moments. They make a great team.

The next layer down is the team at Fourth Estate and beyond who attend to the minutiae of the text. I have been especially lucky to have an incredibly fastidious copy editor in Kit Shepherd. I think authors have a love/hate relationship with their copyeditors but there is no denying that this book would not be anywhere near as readable as it is without Kit's careful work. I was reading David Nicholls *You Are Here* while I was doing my copyedit whose main character is a copyeditor. I really enjoyed how Nicholls channelled all his experiences as an author into the character and it helped me cope with the hard work that this layer involves. At Fourth Estate I benefitted from the careful stewardship of many of the team including Victoria Pullen and Alex Gingell.

The content of the book depends on the encounters I've had over the years with artists who've used these mathematical blueprints in their work, sometimes knowingly, often intuitively. There are too many people to thank in the arts who have invited me into the studio, the rehearsal room, the concert hall and shared their thoughts on the blueprints that are at the heart of this book. Their stories are in the book and I thank them all for allowing a mathematician to jam alongside them and to vicariously live out my fantasies of being a different sort of artist.

The cocoon that encompasses all this process, the outer layer of the whole fractal that makes a book possible, is my family. Shani, Tomer, Magaly and Ina are the backbone that hold me securely as I navigate the challenges of writing each book. I have dedicated this book to my three children. Shani and I have tried to lay out a helpful blueprint for what life has in store but it is beautiful to watch the different ways that they are each realising that blueprint.

Index

Oxford University Images, Joby Sessions

Marcus du Sautoy is the Simonyi Professor for the Public Understanding of Science and professor of mathematics at the University of Oxford as well as a fellow of New College and of the Royal Society. He has presented on numerous radio and TV series, including the BBC's *The Story of Math*, and is also a playwright. He lives in London.

RAISING READERS

Books Build Bright Futures

Thank you for reading this book and for being a reader of books in general. As an author, I am so grateful to share being part of a community of readers with you, and I hope you will join me in passing our love of books on to the next generation of readers.

Did you know that reading for enjoyment is the single biggest predictor of a child's future happiness and success?

More than family circumstances, parents' educational background, or income, reading impacts a child's future academic performance, emotional well-being, communication skills, economic security, ambition, and happiness.

Studies show that kids reading for enjoyment in the US is in rapid decline:

- In 2012, 53% of 9-year-olds read almost every day. Just 10 years later, in 2022, the number had fallen to 39%.
- In 2012, 27% of 13-year-olds read for fun daily. By 2023, that number was just 14%.

Together, we can commit to **Raising Readers** and change this trend. How?

- Read to children in your life daily.
- Model reading as a fun activity.
- Reduce screen time.
- Start a family, school, or community book club.
- Visit bookstores and libraries regularly.
- Listen to audiobooks.
- Read the book before you see the movie.
- Encourage your child to read aloud to a pet or stuffed animal.
- Give books as gifts.
- Donate books to families and communities in need.

BOB1217

Books build bright futures, and **Raising Readers** is our shared responsibility.

For more information, visit **JoinRaisingReaders.com**

Sources: National Endowment for the Arts, National Assessment of Educational Progress, WorldBookDay.org, Nielsen BookData's 2023 "Understanding the Children's Book Consumer"